Clarence Wall

Worlds To Come

AF155946

Clarence Wall

Worlds To Come

A Novel of the Future

JustFiction Edition

Imprint

Any brand names and product names mentioned in this book are subject to trademark, brand or patent protection and are trademarks or registered trademarks of their respective holders. The use of brand names, product names, common names, trade names, product descriptions etc. even without a particular marking in this work is in no way to be construed to mean that such names may be regarded as unrestricted in respect of trademark and brand protection legislation and could thus be used by anyone.

Cover image: www.ingimage.com

Publisher:
JustFiction! Edition
is a trademark of
International Book Market Service Ltd., member of OmniScriptum Publishing Group
17 Meldrum Street, Beau Bassin 71504, Mauritius

Printed at: see last page
ISBN: 978-3-659-70464-2

Copyright © Clarence Wall
Copyright © 2016 International Book Market Service Ltd., member of OmniScriptum Publishing Group
All rights reserved. Beau Bassin 2016

DEDICATION

To my family--my beloved wife and children—whose love and patience have afforded me the circumstances and have helped me to assume the frame of mind needed to write the following tale of the future.

EPIGRAM

There are those that look at things the way they are, and ask "Why?" But I dream of things that never were, and ask "Why not?"

-----Rudyard Kipling

CONTENTS

PAVILIONS: A REVIEW

A well-known critic wrote the following about the film *Proclivities,* whose subject was the proposed 'come-as-you-are' pavilions.

The pavilions, as discussed in documentary style and also dramatized in various scenes, are a kind of amusement park, a kind of Kamikaze one--for adults—where, how shall we describe it, peculiar, perverted, even homicidal impulses are indulged in freely and completely. Some of the film's characters instead go to the pavilions to harm, even kill, others and also risk having the same violent acts inflicted on themselves.

But, in this film, some of those seen in the film, who went to the pavilions, had not paid to be there; they were convicted of violent crimes and were now serving out part or all of their sentence in the pavilions. As they commit acts of violence in these places, they earn points to lessen their sentences, some even earning enough for their release from captivity—if they survive—and some are even hired for well-paid employment in the pavilions. The film's depiction of inhuman behavior in the pavilions reflects the brutal and shocking activities that would take place, according to the author of the best-selling book—that is, if these places were ever actually built and operated. Heaven forbid!

The scenes of drug users shooting and snorting up in the drug pavilions seemed particularly nauseating to me—someone, who in this business for more than 30 years, has viewed and written about the most extreme films of those times. Other scenes depict sexual behavior that run the gamut from the very conventional to the very depraved. And that's not all; oh no! In the shooting gallery pavilion, for a fee, visitors pay for the privilege to drive like maniacs and get away with it unless or until some other driver with a similar mindset and opportunities takes them out and leaves them smeared on the pavement of the pavilion track or impaled on a utility pole. And, as in the killing pavilion, some of those at this pavilion have not paid to be there; I already mentioned criminals who have been sentenced to serve there; but others are law enforcement personnel who have repeatedly violated traffic laws, and other are violators, who blatantly violated laws and drove recklessly, are there --as part of their sentence or for remedial training.

Proclivities is based on the speeches and articles of Congressman Joseph Mufflan, who is currently a candidate for president. If early reports on the film's initial financial intake from ticket sales hold up, Mufflan should reap a bundle, which he can use to finance his presidential campaign and even invest in the building of the real-world pavilions, which he has been talking about for some time. Heaven help us if he ever has an actual opportunity to go through with his plan! But you, the viewers of this film and hearers of Mufflan's idea for the pavilions will ultimately decide if these places ever come to be.

DOING THE RIGHT THING

It was difficult for some to believe that the impetus for the pavilions came from someone who had led such a quite ordinary life—Joseph Mufflan.

As a child, Joseph Mufflan was taught that the guiding principle for one's behavior should be to do the right thing. He grew up in a very religious family, one that believed in obeying the rules, towing the line--or else, facing the consequences. As a young child, attending church regularly was something that became second nature to him, at least during the early part of his life.

One typical Sunday morning, as the members of the Mufflan family hurriedly readied themselves for worship services, young Joe asked his father what, for his family at least, was a shocking question.

"Dad, why do I have to go church?" Joe asked his father.

"Don't you like church?" his father asked, straightening his necktie.

He shook his head. "It's boring," the boy remarked.

"Do you want to go to hell?" His father asked him bluntly. "Well, do you?"

Joe fumbled for an answer and finally blurting out, "No, sir."

"Well, if you don't go to church, that's where you'll end up—in hell," his father assured him.

"But I---" Joe stammered.

"There are no buts, son. Without God, we'll--and I mean all of us --will surely go to hell and deservingly so. So, do you see, son, why we gotta go to church?"

"I guess so" Joe stammered, although he still didn't understand why.

When Joseph was a young boy, simplistic answers like those his father had given him were enough for that time. His father had spoken, and Joe, respecting and trusting him, listened and believed him or, at least obeyed him, and then got ready for church.

Later, as Joe grew up, he would question his father's beliefs. Who said that if I don't go to church that I will go to hell? He asked himself. And, even if it does mean that if I don't go to church, that I will go to hell, shouldn't a person be able to freely decide where he wants to go? To church or to hell.

Like other principles that his father tried to instill in him, going to church was

6

simply the right thing to do--at one time. As he grew, Joe discovered that the right thing to do was subject to change and the belief in what was right and wrong differed from person-to-person and society-to-society.

Sitting in church, his father placed a firm hand on Joe's shoulder in an attempt to stop his son from fidgeting and then re-directed his attention back towards the pulpit, where the minister was beginning his sermon.

"Today's message is from First John, Chapter 5, verses two and three," the minister announced, as he opened his Bible and began begin to read. "'This is how we know that we are the children of God; by loving God and carrying out His commands. This is love for God: to obey his commands.'"

The minister looked up from his text and turned his attention to the congregation. "God loves us; that's what the Bible tells us. It even says that 'God is love.' God loved us so much that he sent his only son to die for us, 'for God so loved the world that he gave his only son.' His son died for our sins. God indeed does love us, but it makes you wonder why. The scripture I just read says that if we love God that we will obey his commandments and do His will.

"Yes. If you truly love God, you'll obey him—with no if, ands, buts, or maybes. You either live according to the will of God or you will willfully disobey him. And if you truly love God, it is your duty to urge others, your brothers and sisters, to obey God.

"How many of you want to go to heaven?" Almost all hands in the congregation went up? "How many of you want to go to hell?" No hands went up. "If you don't help set your brothers and sisters straight and help put them on the right path, guess what? They're going to hell?"

Joe felt like the minister was talking to him personally.

What did rub off on Joe from a sermon like this was that he had a duty to others to help them find salvation. His idea of what salvation was would change over the years, and in the end, each person had to form his own conception of salvation and what they must do to achieve it. As the years passed, he observed that that many decent people were at the mercy of the lawless, depraved, and perverted. He vowed to find a way to protect people from others who sought to do them harm; that sense of duty led Mufflan to form his concept of the pavilions, which would be a means to protect good people from those who were not so benevolent and loving.

7

In listening to other sermons Joe saw his obligation to others. In the one about the good Samaritan, he saw the need for helping others; in the one about the prodigal son, he saw how he might reform those who needed to be rehabilitated if they hoped to achieve some sense of redemption. His idea of duty applied not only to individuals but to entire societies.

As a teenager, Joe began to question the values his family had tried to instill in him. There seemed to be many contradictions in various sets of religious beliefs, such as the accumulation of wealth by many of them, including Christians. If obtaining eternal life in heaven should be the goal of one's life on this planet, why did so many Christians devote so much of their time and effort to accumulate material things? They claimed to trust God, but seemed to seek reliance more on worldly goods and concerns than on spiritual ones. They ignored parts of the Bible that said too much reliance and attachment to material things was a hindrance to reaching heaven.

And why did so many Christians resist sharing their excess material resources with those who had little and were in need? Instead of acting like a Christian, that seemed like a very good way of increasing one's chances of earning a place in hell. So many resisted the idea of getting rid of a lot of that excess material baggage and, at the same time, helping others in need, as Jesus said people should do. Lessening the weight of one's worldly burdens would make one much lightly, so they could more easily float on up to heaven.

And why did so many Christians favor the death penalty and yet oppose abortion? And why did many want to protect the life of unborn fetuses and yet oppose helping children and families in need?

And why did so many Christians try to separate the idea of love for one's fellow man from the idea of Christian charity? Didn't Christ say that if you had more than you needed for yourself that you should share the excess with those in need? Too many Christians seemed to have forgotten that one can't love God and, at the same time, neglect the welfare of ones' fellow human beings.

Mufflan couldn't help observing that church leaders rarely addressed the misuse of wealth, including that used for evil purposes and self-indulgence. It was as if they were afraid of antagonizing the rich who might decide to reduce their contributions to churches if the clergy focused on them. The subject of using some of that excess wealth to reduce suffering seemed to be off limits for many church leaders. And many rich people, who claimed to be Christians, still seemed content to wait for their eternal reward in heaven and, at the same time, while intent on building a heaven on earth for themselves now—and seemed to ignore the fact that they had acquired much of that wealth by exploiting others and making life a living hell for poor and working people.

To Mufflan, what symbolized organized religion was an open hand—not readily offering help to those in need but instead expecting, in reality, to be filled by others. And too often that hand seemed to be directed at those who could least afford to give.

And why did organized religions spend billions of dollars constructing extravagant churches while their leaders preached that this life was only transitory and passing? Why wasn't more of that money being used for those who had real needs now? They seemed to ignore a basic Christian precept that salvation was not achieved though the amassing of wealth. If there was indeed to be a final reckoning, a last judgment, religious leaders were going to have a lot to answer for.

Christ had taught that a church was not a building but a gathering of people. God didn't need those structures; instead he wanted people to improve their attitudes and show their charity in action. Those earthly edifices would surely appear puny and insignificant against the true and eternal heavenly ones.

Church leaders seemed to criticize some sins more than others. But much worse were the sins of omission—sins as a result of not doing what one could for others. Often believers seemed so focused on God that they ignored their fellow men much of the time.

Christians believed that everything was created by God and that people should be caretakers of the earth's resources. They also believed that all of God's creation was good. But, if that were so, why was so much of that creation being used for evil. Perhaps it was because people had subverted the original purpose of God, who intended things to be good, and instead perverted that creation for evil purposes.

And what about religious wars? Oh, that was an easy one, Mufflan thought-- if only one side espoused Christianity. God was obviously on that side; he had to be, didn't he? Would a God allow his followers to be trounced by a bunch of disbelieving heathens? But where did God stand if in a conflict more than one side espoused Christianity? Which side did Christ favor? Whose prayers would He listen to, and whose would he ignore?

Many of those who called themselves Christians seemed to have forgotten about the compassion that Christ taught in the New Testament and instead embraced the legalisms of the Old Testament. It was as if they didn't trust God, and his working through

Jesus and the Holy Spirit, to carry out His will on earth, and, therefore, they felt that they had to focus on earlier parts of the Bible and try to use political institutions to carry out God's plans for Him on earth. They seemed to have forgotten Jesus advice to "Render unto Caesar those things that were Caesar's" (that is, secular and political concerns) and "...unto God those things that were God's" (that is, spiritual concerns). Christians who swore that they trusted God seemed to want to play God, themselves—often through the world of politics and secular government.

No, it wasn't just the Islamic clerics, who tried to play God; it seemed that leaders of all religions felt that they had to be the sole source of spiritual truth and enact what they believed into law—even when that meant discriminating against and shutting out those of other religious persuasions. They seemed to have forgotten what the New Testament said, that no one could live up to the law all of the time, that salvation came solely from faith—not from one's actions. But, judging from the way that some Christians behaved, they seemed to want to play God because, in their pride, they believed that they could do a much better job of carrying out God's plans—than God could. Muslim, Christian, or whatever religious persuasion he or she might be, an adherent, who tried to enact government rules based on the tenets of that religion, in effect, risked making himself into a demigod, even a dictator, and came to believe that they had to rule over everyone else, that based on what they said they believed, that whatever they said or did became the will of God.

It seemed that Christians had betrayed many of the teachings of Christ. Some religious leaders were willing to go to war to carry out Christian principles, as long as they themselves didn't actually have to serve in uniform. But, of course, if one religion believed that it had the absolute truth, then all of the others had to be wrong, or, at least some Christians believed that that had to be the case. Didn't they? Or so they believed. Maybe each religion only had a part of the truth, and various religions had to view various interpretations together in order to assemble a more complete picture of the truth.

Modern Christians wanted to appear as enlightened and tolerant to outsiders, above the realm of politics. They didn't like to be reminded of their own religion's history, especially of its cruelty and abuse during the Middle Ages, a time when Christianity had, by its actions, shown itself as extreme and all-encompassing in the lives of people, as Islamic fundamentalism did in modern times.

Young Mufflan saw the United States as both very religious and also very secular, even irreligious and anti-religious in many ways. He came to realize that the only way that religion could exist in a republic was toleration of all other religions.

So his religious, as well as political and ethical philosophy began to take shape. Mufflan concluded that what was ethically right had to be that which caused the least harm and promoted the most good. Strong followers of some religions insisted on promoting and trying to achieve what, in their own eyes, was right and opposing what was wrong; their efforts to establish what they saw as "the good" could, if embraced, result in conditions that were worse than those they opposed.

Mufflan decided to call his set of philosophical and ethical beliefs "minimalism," that is, doing what would cause the least harm and achieve the most good. He came to the conclusion that what one should do was not simply a case of thou shall or thou shall not. Instead, one must strive to do the most good and cause the least harm.

In high school Mufflan worked on the school newspaper and did a lot of writing on his own. He also participated in debating clubs, and just to test himself, often embraced the opposing or least popular position. In fact, he earned a reputation as a weirdo, kook, even psycho. In college he majored in journalism, and when he graduated, he went to work for a small newspaper. Occasionally, when the paper's regular column writers were not available or an extra piece was needed to fill up an empty space in the paper, Mufflan would be called on to write something. In these columns and articles he addressed a variety of subjects, some of which would lead to his ideas for the pavilions. . Subjects included: crime and punishment, drugs and alcohol use and abuse, sexual behavior, morality and ethics, laws and what should be legal and illegal, fairness in a variety of areas, war and violence, the role and limitations of government, especially as they influenced and even regulated people's lives.

Some of his columns addressed such diverse subjects from the fairness of capital punishment and to how the government raised and used revenue. Mufflan wrote about the abuses of tobacco and alcohol; although these were legal, in some ways, they were more harmful than the kinds of drugs that the government was spending billions of dollars to unsuccessfully keep out them out of the hands of people and to apprehend and prosecute drug traffickers and dealers.

He also wrote about the seeming futility of the government's efforts to outlaw and combat the use of mind-altering substances, which people continued to use, regardless of the costs or what the law prohibited and allowed. He could see that there was a lot of money to be made in the illegal drug trade and spent in law enforcement's efforts to try to control it. Mufflan expressed a belief that if those drugs were legal, their price would drop drastically, since they themselves were not that intrinsically valuable.

Decriminalizing drugs would also save billions of dollars in law enforcement costs—money that could be used for more worthwhile purposes. The fact that these substances were illegal made them expensive, and that seemed to fuel illegal drug trafficking. When money could be made by breaking the law, it seemed that someone was going to try to find a way to do it. Mufflan even suggested that the government could buy up all of the illegal drugs for a small percentage of what it now paid in law enforcement expenditures and then give or sell them cheaply to drug users in some very tightly-controlled way and put the dealers out of business. He believed that such a move would reduce the number of crimes related to the drug trade and addicts' efforts to obtain money to obtain drugs to feed their habit.

In other columns he wrote about gambling, how it was detrimental to the work ethic. He came down on government sponsored-gambling; he believed that government leaders, including legislators, were too cowardly to try to collect what the government needed to operate. He also contended that the government was too wasteful in using the revenue it did take in and too vague in informing taxpayers how the money it collected was being spent. But it seemed that a lot of people were far more willing to throw away their money gambling in lotteries and at casinos than in paying their fair share of taxes, regardless of how worthwhile the programs government revenue was being spent on.

He even wrote a column comparing investing, such as that in the stock market and real estate, with gambling. He concluded that one of their main differences was that gambling was far more honest than conventional investing and much better regulated and monitored. To Mufflan, there seemed to be more crooks in business investment firms than in the casino business, and many insiders seemed to be trying to make money off of the ineptitude of other investors than from an honest return on the investments themselves.

In other columns he addressed the disparity in the distribution of the world's wealth, most of it now in the hands of relatively few people. Were the excesses of investing the mark of a Christian nation, one that had resulted in a few very rich and multitudes of poor people? If money was used wisely, it could solve many, if not most, of the problems of the world. Instead, it had been used to increase suffering among the less fortunate. Even government policy favored the wealthy and exploited the poor. Material resources, which were capable of being used to achieve great good, were often being used for evil and to oppress others.

Mufflan also wrote about fairness and promoted the concept of the need for a level playing field in society. He contended that his country may have once provided a better environment for actually promoting opportunity for all. Did it ever exist, he asked himself. Maybe, not, but the idea did. Governments seemed to be expending a lot of effort to promote the myth that it still did, while, in reality, its policies made the economic and political playing field more uneven than ever.

Mufflan even addressed the subject of driving and traffic laws. He saw the rampant and deliberate violation of laws and law enforcement's inconsistency in enforcing standards and its extreme reluctance to enforce existing laws as the worse crime problem in the nation, maybe even in the world. The situation, in his eyes, was like that in the Wild West, in that drivers pretty much did what they wanted to, and few paid-government officials seemed to voice any serious objections until someone was killed or seriously injured or there was significant property damage.

The few drivers who did try to obey the laws were at the mercy of the speeders and tailgaters and could expect little help from the police, many of whom repeatedly violated the very laws they had sworn to enforce. In many areas, there were situations where numerous laws had been enacted to make driving safer, to protect people and property, and most of those laws were largely ignored or enforced selectively. Having someone with that kind of lax attitude behind the wheel of a motor vehicle turned it into a deadly weapon and the driver into a potential killer and other drivers into potential victims and targets. Mufflan viewed reckless driving as assault, at worse, and as negligent, at best, and believed that such behavior should be punished severely. Calling a collision an accident when a driver had deliberately placed others and himself in danger by his carelessness was a perversion of the language, as well as the driving privilege. Law enforcement too largely ignored, even deliberately violated, the very laws they were supposed to be enforcing. If there was only some way to isolate those lawless drivers all together on the roads and let them kill each other off, then safe motoring might finally become a reality.

Some of Mufflan's columns questioned what the government's role should be in legislating human behavior. Quite often those laws that had been enacted didn't seem to be very effective; so why keep passing them, anyway?

Mufflan also addressed various sexual issues in a number of his columns. In one, he advocated the legalization of prostitution. In society, people, in effect, often sold their souls to get ahead materially; so, why not profit by selling or renting out one's own body, if that's all one had to barter with? He had written that the government should not attempt to regulate sexual behavior, including homosexuality, as long as that conduct did not hurt others. However, he did not take a position on homosexual marriages. He reasoned that most people made a mess of being married anyway and corrupted a basically decent arrangement by not living up to a few simple but essential standards. Why should heterosexual married people be telling homosexuals that they couldn't get married to other homosexuals when most heterosexual marriages failed? And the fact that most of the male customers, who frequented prostitutes, were married spoke very critically about the inadequacies of many marriages.

His opinions on pornography in his columns could be summed up as this: what a waste! It, like too many other concerns, was unhealthily motivated by money rather than providing decent services and products at a fair price. Pornography was produced to make money for those who produced and sold it and to try to make ordinary people feel that their own sexual lives were inadequate, as well as to market a lot of other commercial products and services. It sadly illustrated the point that people would spend their hard-earned money for trashy purposes and products rather than on worthwhile products, services, and causes.

Mufflan wrote columns that addressed the hypocrisy and unfairness in the country's various legal systems and standards of punishment. There were at least two tiers—one for petty criminals, the poor, and minorities—and another for the rich and influential. White collar criminals were often punished with a slap on the wrist while the less influential were incarcerated for violations involving very small amounts of monetary damages and for longer periods of time. He thought that sentencing guidelines should be based on his minimalist guidelines. He believed that an offender who killed or injured others was inflicting far less harm than someone who stole millions or billions of dollars by cooking the account books and, in the process, ruining the lives of thousands or even millions of people. He believed that a determining principle in punishing an offender should the harm the offender's actions had caused. He also believed that corporations and government agencies that were charged with crimes should also try the individuals whose actions and harmful decisions had caused the damage or harm. After all, corporations or agencies were not the cause of crimes; individuals within them were.

In some of his columns Mufflan addressed the subject of crime and how society should deal with it. Building more jails and prisons seemed to be the only official solution for dealing with crime—hardly an effective remedy. In fact, the amount of money being spent to house one criminal was more than the income earned by the average working person. Again, his concluding reaction was: what a waste! The problem of crime had gotten out of hand, and continuing to spend additional amount of money to resolve the problem didn't seem like a very realistic means for fixing the problem. If only there was a way to isolate criminals so that they could prey only on other criminals, instead of on the decent, innocent, and harmless. Isn't that what the rich try to do, Mufflan asked himself—isolate themselves from the criminals and the "unwashed" masses? Of course, many of the rich were themselves criminals, but of a different kind: respectable and usually not the kind that were ever imprisoned or jailed.

In one column he examined the effectiveness of law enforcement itself. A lot of people didn't trust their governing officials, and many others viewed them as unable to even protect themselves--without overreacting—too frequently including killing unarmed citizens while going after minor offenders. Whatever happened to the principle of reasonable force, he asked himself. Mufflan even suggested that, since people in many instances took the law into their own hands, maybe there should be vigilante groups organized at local levels—more citizen participation and fewer full-time paid law enforcement personnel. After all, were the billions of dollars being spent to pay those officials really worth it; were they really preventing crime, which is what a peace officer was supposed to do?--not simply deal with crimes after they had been committed.

Mufflan tackled the subject of gun ownership in other columns. He believed that his findings conclusively revealed that more citizens owning guns would not reduce the amount of crime, but, in fact, would increase it. Fist fights and arguments could quickly escalate into shootings when firearms were readily accessible. Law-abiding people who owned guns rarely used them to protect themselves against criminals; they usually killed those closest to themselves, accidently or in fits of anger. Besides, the second amendment, if one read it correctly, prescribed that gun-ownership was a requisite of membership in a militia to protect ones community. In other words, if people wanted to own guns, they should join their local National Guard unit.

Needless to say, many of the subjects of his columns stirred up a great deal of controversy. He lost many jobs as a journalist, but always seemed to find other ones quickly. At first, he was welcomed by many newspapers and magazines, but after writing a number of controversial columns, he quickly perceived that he had outlived his welcome and was unceremoniously asked to leave.

After several years of moving from place-to-place and job-to-job, Mufflan discovered that he had accumulated a considerable number of columns and articles on a variety of subjects and decided to try to publish several collections of his writings in book form. It took Mufflan months to obtain permission to use the columns from all of the places he had been employed. The books were instant best sellers and were reviewed in a number of highly respected publications. He appeared on numerous television and radio programs to plug his books and discuss some of his other ideas. He then decided to write his columns, on his own, independently of any one newspaper or magazine, and worked out arrangements so that a variety of newspapers paid a fee to receive his columns and an additional fee for publishing them. He was also hired to do his own regularly scheduled public affairs program on both TV and radio.

As Mufflan began to acquire wealth, he experienced tinges of guilt and wondered if he had betrayed what he had believed in. He lived simply and donated a lot of his earnings to charity. He carefully invested his extra income and earned even more, but continued to donate most of his earnings to what he viewed as worthy causes.

In the process of writing numerous columns, Joe Mufflan had found himself formulating a way for dealing with various crimes and vices—by isolating the commission of those offenses or indulgences from the rest of society. He reasoned that there had to be better ways than simply imprisoning offenders and enforcing prohibitions of certain vices that had been outlawed but that could not be prohibited very thoroughly throughout society. He envisioned places, which he called pavilions, where vices and even crimes could be practiced without polluting the rest of society. Criminals would also be sent to these facilities to serve their punishments, and, while there, their minds would be kept occupied by worries and concerns about how they could survive when they were up against other offenders who had few qualms about inflicted violence on others, and, in the process of brutal confrontations, many would be killed off—many after taking other offenders with them. Also non-criminals, that is, those who wanted to practice certain vices would pay to go to the pavilions where they could freely practice them.

Mufflan wrote a series of articles on the pavilions, describing how they could improve society, and other articles in even greater detail on what would take place in the various pavilions. Later in another book he laid out a plan in much more detail about how the pavilions could be used to accommodate various vices and crimes. At the heart of the idea of the pavilions was the principle of minimalism. Using this concept, he addressed the question: how much was currently being spent to enforce certain laws and standards of morality? He came to the conclusion that a government did not have the resources to prohibit or even protect people from all of the vices that they may have wanted to indulge in or avoid or, possibly, even the right to try to prohibit or control.

He believed that establishing the pavilions might be way to bring about a better society. So, many vices and crimes that plagued society could be isolated to designated places, meaning, in the pavilions, so that they did not affect and harm the rest of the people. To do this, he would have to work to enact some new laws, and to do this, he entered the world of elected politics.

Joseph Mufflan ran for Congress and was elected and re-elected twice. During his third term he was elected as Speaker of the House, and near the end of that term, he had an associate representative introduce a bill for establishing the pavilions on an experimental basis. The bill was subject to much debate and was still in committee during the campaign, Mufflan's second run for president of the United States.

--3--
...NOT EXACTLY A LANDSLIDE

The evening of the general election President Nathan Meeker watched the returns from the Oval Office of the White House as they came in on TV. He nervously sipped from the glass of spirits as his associate nodded in anticipation of another victory for the president; early but mixed returns were a very hopeful sign that President Meeker might squeak by this one last time and retain the presidency for a second term. "I wish Judy was here to help me get through this night," Meeker remarked. Judith, Judy for short, Meeker's wife, had succumbed to cancer six months earlier.

Meeker's main opponent, Joseph Mufflan, had voted near his home earlier that day. He had already shown considerable strength in a number of key states. But the lead between the incumbent and the contender shifted back and forth a number of times throughout the night, leaving the outcome in doubt for much of the evening.

Later, news reports would indicate that President Meeker had been drinking brandy, the quantity of his intake increasing as the night wore on and his chances of retaining his office diminished. A commentator on one of those off-beat radio stations had sarcastically remarked, "If I was getting creamed at the polls like he was (and he indeed was in a number of areas), I'd get plastered too."

That evening, at his home, Joseph Mufflan laughed when he heard a TV journalist refer to him as the contender, recalling a familiar movie line and now voicing the line, "I could have been someone. I could have been a contender."

"You've always been a champ in my book," his wife Marie replied, as she kissed him on the cheek.

"Thanks for the encouragement," he muttered.

"I'm hitting the sack," she announced. "Don't stay up too late, dear. Tomorrow is going to be a big day—either way."

"Yeah, I won't. I'm a little hyped up. After all, how many times does an average Joe like myself get a chance to run for President."

"Average Joe? In my book, you're way ahead of the pack," she told him. "Good night, dear."

So it seemed that Joseph Mufflan, the candidate of the other major political party and a fifth-term Congressman, was holding his own against an incumbent President. Who would have dreamed that someone with such radical views and ideas would come this close to being elected President of the United States? Not that many, but a lot of politicians as well as ordinary people were talking about Mufflan's ideas—a lot.

A significant part, perhaps the greatest, of Mufflan's program was his proposal for establishing and operating the "come-as-you are" pavilions. These places were intended to serve as outlets for channeling idiosyncratic and socially negative impulses and behavior into and away from general society. Mufflan told voters that he believed that these pavilions would not only be places where a lot of bad behavior could be vented but also promised that the government's share of income from the pavilions would bring vast amounts of revenue into the public coffers. This new source of revenue could reduce taxes, reduce the size of the government deficit, payoff the national debt, and fund new programs; the government could even rebate a part of the money received at the pavilions to taxpayers.

Muffland believed that providing jobs for ex-convicts in the pavilions would not only reduce the amount of crime but also the number of people who needed to be incarcerated in taxpayer-funded penal facilities. Although many people considered the kind of practices and behavior that would take place in the proposed pavilions to be undesirable, even perverted, Mufflan had insisted that whether one deemed them negative or positive, to be pursued or abhorred, that evaluation was dependent on one's point of view and value system. He also insisted that the government was responsible for fostering and promoting conditions in which the people could, with minimum limitations, live and do as they pleased—not for legislating or enforcing a specific code of morality.

After years of observing the behavior of humans and official attempts to discipline and even punish them for what they did, Mufflan had concluded that many laws, even those enacted with the best of intentions, were unenforceable within the constraints of government's limited financial and manpower resources. Some of those laws involved

19

behavior in an often vaguely-defined category called "victimless crimes. From that distinction arose the question: should the authority of a state be used to try to prosecute those who committed them–when there were no victims? "If victims could not be defined for these so-called "victimless crimes," should the government still try to enforce them?

If the pavilions were to become a reality, as Mufflan hoped they would be one day, he hoped that he could attain the position of president and be able to sign a bill into law authorizing the construction and operation of the pavilions. But even as a candidate, he had his allies and supporters in Congress, one being Congressmen Phillip Anderson. Even while Mufflan was a candidate for President, Anderson introduced a bill in the House of Representatives concerning the establishing a limited number of pavilions on an experimental basis.

The TV networks interspersed the coverage of the election returns with footage of speeches of the two major candidates, many being about Mufflan's advocacy for the pavilions and President Meeker's response and position on related issues.

Mufflan had made a lot of speeches, not all of them related to the pavilions. "I served time in the military," he had said, on the floor of the House of Representatives, when he voiced his views on a proposed military bill.

"Makes the military should like prison, doesn't it?" a colleague muttered.

Mufflan chuckled and continue. "I guess it does, but I won't get into that here. The country shouldn't go to war unless its people are behind it and are willing to sacrifice to fight and win it. People die in wars, they are maimed; their minds, their whole outlook on life are changed by war, including PTSD, that is, post-traumatic-stress-disorder. My dad told me that they used to call that combat fatigue."

At times, when speaking on the floor of the House of Representatives, Mufflan focused on more general concerns, some of them related to the pavilions by implication. "This distinguished legislative body should not enact laws, including treaties, where the authority behind them isn't willing to give its all, including risking the lives of law enforcement and military personnel to enforce them. We have to ask ourselves this question: Are we willing to go to whatever lengths that are required to enforce laws or treaties, I mean in terms of committing and, if necessary, risking and sacrificing lives and resources to enforce them?"

In a serious of speeches, delivered at various times, locations, situations, and occasions, Mufflan discussed his ideas for the proposed pavilions, including how they could be used to ameliorate some of the negative conditions in the country, including crime and vice. He repeatedly described how such places could bring in large amounts of additional revenue to the government and thus reduce the amount of taxes it needed to collect and also reduce the amount of crime and the current prison population.

In one speech, Mufflan discussed the lack of enforcement of traffic laws.

"Take traffic laws, for instance," he told the audience

"You forgot the 'please,'" someone in the audience blurted out."

Mufflan chuckled. "Take my wife, please," he chuckled, as he muttered under his breath." Few in the audience heard the remark. "Where was I, oh, yes, take, traffic laws."

"You can take all of those damned laws and shove them...." someone else yelled out.

The host of this meeting came to the podium and banged his gavel for order

"No, he's right," Mufflan interceded. "Yes. He's right. Most traffic laws are unenforceable and unenforced. I too deplore the way many people drive--recklessly, flaunting traffic laws with impunity, endangering the lives of others, as well as their own. And they keep doing it, and you know why? Because the odds of not getting caught are in their favor. And are most member of our police force willing to risk their own lives and jeopardize the lives of other motorists and pedestrians to try to apprehend speeders? But we have to get those speeders and other violators off the roads." He noticed a raised hand in the audience. "Yes," he said, pointing to the questioner.

"So what do you propose to do about it?" the questioner asked.

"I'm really glad you asked that question," Mufflan answered. "One of the pavilions I intend to propose will deal with unsafe drivers and traffic violators. I'll say more on that subject when the details of my plan for this pavilion have been worked out. Let me just say this: If the necessary laws can we passed, then many of those who violate the law will be sentenced to a pavilion where they can learn safe driving techniques or else suffer the consequences of their negligence. At this pavilion, they can drive any way they want, but unlike on our roads and highways now–that is, with limited repercussions to them--there they will face the consequences, which will be very severe, sudden, and immediate.

When Mufflan ran for Congress, many, if not most, people regarded him as a kook. Oh, sure, he had some funny ideas, not the ha-ha kind, but funny in the sense that they were just too ludicrous to be taken seriously by most. In the minds of a lot of voters, and even in those who didn't usually vote, candidates would say almost anything to get elected. Even if the policies Mufflan proposed might have seemed were outrageous, many people had voted for him anyway, perhaps out of curiosity; at least his ideas were different from the run-of-the mill candidate. And even more voters were also lured by his promise of bringing in additional revenue and reducing taxes from the pavilions which Mufflan had proposed.

In the previous general election Mufflan had run for President as an independent on the ballot in 23 states but didn't win a single electoral vote. President Meeker, a conservative and his party's nominee, had won almost all of the electoral vote but had barely squeaked by with a plurality of the popular vote. This time Mufflan was not running as an independent; he didn't have to: the other major national party had unexpectedly offered him its nomination, and Mufflan had readily accepted it. After the nominating conventions were over, President Meeker had been running scared because, with the machinery of a major party behind him, Mufflan just might stand a chance.

On a number of televised public information programs, Mufflan tried to lay out his ideas regarding the pavilions. When opponents appeared in the same forums, that wasn't easy because they continually interrupted to interject their objections before he could fully present his ideas.

On one such program, Mufflan had said, "I propose establishing a series of facilities; I call them 'come-as-you-are' pavilions. For a fee, those with certain, shall we say, proclivities, can go there and indulge in their chosen behavior without hurting others—in society, that is--without airing those--I hate to say--disgusting practices in polite society. Those acts--and I do concede, as a number of my opponents have--that many of them are very distasteful. But I ask you this: just because you or I don't want to engage in those activities, does that give me or anyone the right to keep others for doing so?

"It is the duty of governments to protect its citizens from abuses by other citizens and by the government, as well. And, those very necessary laws to secure those ends will be enforced, and additional ones will be proposed, as needed. But how much of a government's effort should be devoted to controlling certain types of behavior of others? Governments at the national, state, and local levels spend billions of dollars each year to enforce laws that punish those people who want to do what some consider disgusting, and, in many or, even, most cases, those laws do not achieve their desired ends.

So why not let those people pay for their indulgences—and I do mean pay—and use part of that money to finance more important concerns, and not use the resources of the government to foist someone's code of morality on others. After all, who is to decide what should and should not be the acceptable standard in the first place? There have been and still are so many codes and standards that one may live by. But the government should not try to control the purity and impurity of human behavior; so why not try to channel it into the kinds of places I propose and use the fees raised to finance truly worthy causes?"

Mufflan's motives for raising money from other sources than taxes appealed to many people. Opinion polls showed that much of his appeal to supporters came from the prospect of filling government coffers without raising taxes.

In defense of his proposal for the pavilions, Mufflan had said, "In attempting to enforce laws against victimless crimes and other indulgences, many other concerns are being neglected. Because vast sums of money have to be spent on law enforcement, including incarceration, many other important areas have been neglected for too long.

"Should governments be responsible for controlling the morality of its people? I know that a lot of people will say that governments are responsible for establishing and enforcing moral standards, but whose code of morality is to be used to set the standard? Instead, let's channel immoral behavior–if one chooses to call it that--into the pavilions.

"Let's channel some of the kind of crimes that can result in government incarceration into punishment and rehabilitation in the pavilions. These will serve as outlets for offenders and would-be offenders. The pavilions will save money in the costs of fighting crime, and some of the money taken in at the pavilions can be used to deal with the most serious types of crimes.

"And who will speak up for the underdog, the little guy, who is constantly being screwed or neglected by the government and by society in general. I hope that I may be his spokesman. I will try to be if I am elected president.

Mufflan believed that the pavilions could become an important source for rehabilitating major and minor criminals, including officials from all levels of government, who had abused their power and committed illegal and unethical acts, and that including law enforcement personnel and a lot of white-collar criminals. If they could not be reformed in the pavilions, that place could instead be a source of punishment for those types of offenders.

Would-be murderers, burglars, rapists, and other types of violent offenders— instead of preying on victims within society-- could go or be sent to the pavilions and kill, maim, rob, rape, etc. And there would-be victims would be those who had committed similar violent acts or who had voluntarily paid to be there. For example, rapists, who would go to pavilions to rape, would find willing victims who paid to go there for the privilege of being raped. But those same rapists might find themselves at the mercy of other rapists, and who knows what sex their rapist may be? Mufflan reasoned that when a person strayed from the normal rules, of society, he or she shouldn't be able to cry foul if they find themselves victims of others who stray too.

Major criminals, who were apprehended and convicted, could be sentenced to the pavilions as punishment, where they had a very good chance of being killed or badly injured by even more brutal criminals then themselves—kind of like the gladiators of ancient Rome, who even as they faced the very real prospect of being killed, sought to win glory and even their freedom by killing others in combat, and, in a way, they were doing a public service by killing or incapacitating other violent criminals. The pavilions could save vast amounts of money in costs for incarcerating violent offenders. In fact, a convicted criminal being sentenced to serve time in a pavilion would at first view time served there as a virtual death sentence. But, on second thought, the pavilions did offer a choice to being executed or interned for years, even for life; there was even a chance, if they could reform themselves in the pavilions, that they could gain long-term employment there. So, there were opportunities for rehabilitation in the pavilions, provided that those sentenced to them survived and could demonstrate that they had undergone some kind of internal change of attitude and outlook and were willing to use what they had had sadly learned from the consequences of breaking the law against other lawbreakers. Mufflan hoped that some of the more vicious, but promising, offenders could be reformed in order to staff some of the pavilions.

Mufflan even recorded promos to communicate his ideas about the pavilions as a possible way to reform criminals and as a system for dealing with those who, society viewed, as unwilling to shape up or fall inline. Looking directly into the camera, he said, "Are you tired of being the victims of crime--at least twice; first, for the losses and harm their crimes have inflicted; and, second, for having to pay to support them in prisons for years, even for life? I know I am. So why should decent citizens have to expend valuable resources that could be used for much worthier purposes than to lock up convicted criminals safely away from us –those who have preyed on decent members of society. These offenders: they owe a debt to society, and I'm going to do all that is within my power–in Congress or, if you elect me–as you president–to see that they pay that debt. My name in Joseph Mufflan, and I will propose that criminal offenders be sent to the pavilions where, in one way or another, they can begin to pay back their debt to society– to you, the decent people of this country."

Mufflan also spoke out against programs to deal with illegal drugs. "Drug interdiction programs have not worked. We're losing the so-called 'war on drugs,' and a lot of people are getting hurt in the process. Why must innocent people get mugged, beaten, even killed, so that junkies can get that next fix?

"The pavilions will provide places where people can go and indulge in their drugs of choice–away from society–on the premises of the pavilions. This isolation of drug use will reduce the negative effects on the rest of society. The use of these substances will be restricted to the pavilion; they will not be taken out and possibly be resold. There will be detectors near the exits of the pavilions, which will be able to detect any drugs someone might try to sneak out. And if would-be smugglers are caught, they will prosecuted and punished, fined and sentenced to service in one of the pavilions or at least banned from them for a period of time."

Mufflan often discussed his ideas concerning sexual activity in the pavilions. "I know that many of you have misgivings about the pavilions as places where intimate activities will take place. I understand your apprehension; after all, this is a very personal subject. In the pavilions there will be prostitution, massage--catering to both male and female–and many type of intimate behavior, including heterosexual, homosexual, and other varieties.

"Those workers who perform massage and sexual acts in the pavilions will be licensed and will undergo periodic health checks. Those who want to engage in various types of sexual activities are likely to find what they seek in one of the pavilions. Some of you will probably find what I have said as shocking, but let me say something in the defense about those practices: Those acts I have described and many others now go on everywhere, and I'm not talking merely about what takes place in intimate relationships. I'm referring to those that take place in far less discrete circumstances. But when the pavilions become operational, people can feel free to practice them in the safety of the pavilions, without disrupting the rest of society."

After one of his speeches, when Muffan was taking questions from the audience, someone asked the question, "So how much will the government have to spend to build and run these pavilions of yours?"

Mufflan smiled and then made his reply. "Not one red cent. The pavilions will be built, maintained, and operated using private funds. Those who provide the funding and operate the pavilions will make a bundle, but a significant portion of the revenue earned from the operation of the pavilions will go into government coffers. I am already negotiating with a number of investors who have expressed an interest in providing capital to build and operate the pavilions.

"Thank you for coming," Mufflan told the audience. "In closing, let me say this: So, what I'm proposing are places to channelcertain human impulses and to protect law abiding citizens and society in general—by providing discrete, isolated outlets. Those are the pavilions. Thank you for your attention."

In the wee hours of election night, returnss and computer projections indicated that Mufflan would win a narrow victory as president, hardly the landslide that some supporters had hoped for.

FROM THE SOAPBOX

Mufflan hired people to develop plans for building the pavilions, conferring with them on an almost daily basis to discuss his ideas, to illicit their feedback, and to review what had been done and what still needed to be accomplished.

Mufflan's pavilion committee concluded that it would be easier and cheaper to begin by establishing the substances and sex pavilions first because these required a minimum of capital and resources. Also the government had a large stash of confiscated drugs that could be used for the drug pavilions. Mufflan quickly drafted directives to have these made available to pavilion officials.

Mufflan realized that establishing certain pavilions might require the passage of additional laws sanctioning certain activities in them. There would also have to be regulations for raising money to build and operate the pavilions and also for handling the funds of those who invested in the pavilions and also for distributing the monies received from the operation of the pavilions.

In spite of the initial laws that authorized the building and operating the pavilions, each specific pavilion would require enactment of more specific laws permitting certain activities within that pavilion and prohibiting or limiting them in society outside the pavilions.

Mufflan continued to make speeches on the pavilions, when his presidential duties permitted. And he was severely criticized for this by those who protested that "they wanted a president--not a drug pusher or a pimp." Needless to say, he took such criticism and even worse castigation with a grain of salt. In one speech President Mufflan's tried to explain why he spoke so strongly in favor of the pavilions. "A fundamental principle in the enactment and enforcement of laws must be consideration of what should be the role of government in the lives of people.

"I am speaking of what have been labeled by some as victimless crimes. I am not unaware that there are those who want the government to outlaw, pursue, prosecute, and even imprison those who do certain distasteful things. I cannot in good conscience allow that to happen, and I will do all that's in my power to prevent it. And, in many instances, for the government to continue to commit vast amounts of money and manpower to go after victim-less crime is a tremendous waste of tax payer money.

"I have consulted authorities on this question and have found that it is not expedient for the government to totally outlaw certain practices–control, regulate, yes– but completely outlaw-- no; that has been a tremendous waste of the government resources. Many of you know what happened during the period of prohibition during the last century when our government enacted laws to prohibit the manufacture, sale, and consumption of alcoholic beverages in the country. That period was characterized by the worst period of crime in our history. And why was that so?

"During prohibition, people wanted to drink alcohol and were willing to pay to get it. That demand created a vast underground market, which criminal elements tried to profit from; there was a large increase in the number of criminals who were willing to break the law to supply the country with booze. The government spent exorbitant sums of money in its effort to enforce the laws against drinking, but that program was a dismal failure, and those laws were finally repealed. I don't think it is farfetched to say that the laws against booze created a lot of crime. Similar prohibitions are now resulting in more criminal activity; and too many government restrictions can increase and has increased the rate and number of crimes.

"When the government enacts these types of laws that result in a scarcity of products and services, demand for them increases, driving prices up. In fact, the increased opportunities to earn vast amounts of money makes dealing in those restricted items more lucrative than before they were declared illegal, and people turn to crime to scoop up some of that money. Yes, there is even a system of supply and demand for criminal elements, and they employ it in deciding how to supply illegal needs—the more money to be made from something, the more criminals that there will be who will go after it.

"Fighting these types of crimes, such as drugs, prostitution, etc., actually increases the number of crimes committed and the number of criminals who commit them. When the government tries to crack down on certain products and practices, it's like pouring gasoline on an open flame. Fighting a war against victimless crimes--in a half-ass way-- will not eliminate crime. For example, if the government were really sincere in cracking down on outlawing illegal drugs, it must be willing, if need be, to arrest and even kill providers, suppliers, and users; now those people get off with a slap on the wrist. If a government or nation is serious about stopping drug use, it must be willing to actually make war on countries that produce, supply, and participate in drug trafficking. Making trite pledges that the government is declaring war on something--and yet fighting in a half-ass way--portrays the government as totally ignorant of what a real war involves and the costs and sacrifices that are required to fight one.

"There would be less crime, if the government did not try to outlaw certain practices and behavior. My administration will not do that; instead, it will make the proceeds of some vices benefit others, even those who strongly oppose them.

"For example, once the pavilions are open and in operation, there will be fines for prostitution outside of the pavilions, when one can readily partake of the same vice in the pavilions with impunity. On the outside, that fine might be $500 per sex act for both the customer and to the prostitute.

"The goal of my administration, in dealing with certain practices that now take a toll on society at large, is not to outlaw them. No, instead, we will tax them out of business in society and confine certain behavior to the pavilions.

"In the pavilions, one will be able to obtain booze, tobacco, drugs, sex, and can indulge in these and other activities and substances and super cheaply—in fact, cheaper than these items and activities now cost in society, now obtained through the black market. But when the use of these indulgences is confined to the pavilions, society will be insulated from most of these sleazy activities.

"One side of me would like to outlaw many of these vices, but instead their practice that will be confined to the pavilions. Outlawing them-- politically that wouldn't work because as was the case during the time when there was prohibition against alcohol, it just would not work now. I do not believe that I could get laws passed that prohibited certain activities and substances. Therefore, I will try to go in the back door, that is, by using the pavilions as a means of reducing the impact of certain behavior on society.

"Many people will fight for the right to partake of these vices; now they will be able to indulge in many of them in the pavilions, but, unfortunately, only a very few people will ever go out of their way and lift a finger or fight for decent causes.

"Another of my goals is to eliminate or at least reduce the use of tobacco and alcoholic products; these have been proven conclusively to have caused great harm to individuals, to health, to disrupting families, and to reducing productivity. I would also like to commission some detailed studies on comparing illegal drugs with legal ones, including tobacco and alcohol, especially in terms of the damage they have caused and how they have disrupted society. The use of tobacco not only sickens and even kills those who use it, but it harms others who must breathe in second-hand smoke, clean up the litter of discarded butts and ashes, sickens others who must contend with smokers' smelly clothes and bodies reeking with tobacco residue. And we should not forget the fires and loss of lives and damage caused by careless smokers, including that done to our environment

"One way to reduce tobacco use may be to impose heavier taxes and limit places in public where smokers may partake of their vice. The cost of tobacco and alcoholic products is not limited to the purchase price, but health problems and lost productivity, and therefore users of those products should bear much of the costs of side effects resulting from their use.

"If the government bought up most or all of tobacco and booze, it could destroy what it could not sell or dispense at the pavilions. Although the use of some substances may one day be confined to places such as the pavilions, no such restriction will be made on tobacco and alcohol products at this time, pending the results of further studies; I will try to enact laws for higher taxes on these harmful products."

The pavilions would offer a myriad of diversions visitors could choose from, including the drug pavilion.

Mufflan realized that the manner in which the pavilions were publicized was very important to their success. He had some of his staff design detailed plans for advertising and publicizing the pavilions. He wanted this information to be honest, but also as enticing and appealing as possible to certain elements of the population. For instance, Mufflan tasked his staff to come up with a catchy name for the Drug pavilions; he could see that simply labeling that pavilions as a place, where people smoked, snorted, and shot up, as the Drug pavilion was not a very a catchy name for it. His assistants tentatively labeled the drug pavilion as "the High Adventure Pavilion."

Mufflan's plan for supplying the Drug pavilion was that the U.S. government would negotiate with foreign governments to buy up large quantities of various drugs that those countries produced and at much higher prices than those countries received via the illicit drug trade; independent drug merchants couldn't compete with the government's large purchases. (If the U.S. had decided to simply destroy all of those drugs, the money spent to purchase them would still have been far less than the amount that had been and that was being spent trying to enforce drug laws and pursue and prosecute offenders.) Even the strongest critics of the proposed pavilions had to admit that Mufflan's purchase plan was a great way to keep illicit drugs out of the hands of addicts out in society. The government would save a vast amount of money in drug law enforcement costs.

Visitors to the Drug pavilion would pay a set fee, and for this they could use as much of the featured drugs as they wanted, but they were not allowed to take any of the drugs out of the pavilions. The cost of the drugs consumed by the typical visitor represented only a small portion of the entrance fee a customer had paid. The government's share of the income from the sale of drugs would net a vast amount of money, which could be used to fund other programs, including treatment for those who sought help in breaking their addiction.

Those hooked on drugs would no longer have to steal, rob, injure, or kill innocent people to get enough money to buy that next fix. They could get all they wanted easily and inexpensively as long as they were willing to use them at the pavilions—and at a much lower cost than they had to pay on the street.

In spite of the almost unlimited quantity and types of mind-altering substances that could be obtained in the pavilions, out in society drug laws were still strictly enforced. Those apprehended and convicted for violation of those laws still had to be dealt with. But instead of sentencing them to incarceration, often for considerate periods of time, many were sentenced to service and punishment in the drug pavilions. Non-violent offenders could even be sentenced to work there, and many settled down to a more regular routine, where they could obtain free drugs, but on a controlled basis—after they had pulled their work shifts and were off duty. For violent drug offenders, the punishment was much more severe. Because they were not only harming themselves and others by their drug use and violence to obtain drugs, they were also dangerous to others who happened to be in their way of obtaining them. Violent drug offenders would be sentenced to the homicide pavilion, where there would be no drugs available, and where they would have fight almost constantly to survive; needless to say, most of these people would not survive there for very long.

Once the pavilion were in full operation, the small quantity of drugs that would still reach the streets of the U.S. and that would be sold on the black market would have to be sold at exorbitant prices to make up for the significantly reduced supply that suppliers could obtain—at much higher prices than most users and addicts could afford to pay. The lack of customers, Mufflan hoped, would soon put most of the illegal drug trade out of business.

In addition to the drug pavilions, the following pavilions were planned to soon welcome visitors:

The Intimacy Pavilions (also referred to as the Meat Market); this included the massage, sex of various types and with different genders, and the Turkish Delight option:

The Shooting Gallery (also called the Wild Drivers pavilion);

The Homicide Pavilion;

The homicide pavilion, which would be open to those who wanted to die, that is, to have someone kill them, rather than end their own life in an act of suicide; they could pay to go to the homicide pavilion and be quickly, if not painlessly, killed. Before they would be allowed to enter this pavilion, they were required to pay for a life insurance policy to provide for their survivors and, if there were none designated, after their death, the benefits would be paid to the government and used to fund various social programs.

Several major insurance companies were not ready or sure how to deal with deaths and injuries that would took place in the pavilions; that was a matter that still had to be resolved. Mufflan hoped that with the pavilions as an alternative that suicide would no longer be a desperate act but would be a kind of sacrifice that even benefited others.

Those entering the heart of the tentatively-titled the killing pavilions wouldn't know what to expect—they might be clubbed to death, shot, stabbed, electrocuted. Who knows? Mufflan might have joked: "the shadow knows, but he's wasn't telling." Most of the usual visitors to the pavilions would eventually become victims of violence, if they stayed in the pavilions long enough. Undoubtedly there would be skilled and experienced killers, who would come there to kill and inflict injury on as many people as quickly as possible, but even the most experience killers could became victims of amateurs, and these people too had to be ready to defend themselves against other assailants with the same violent intentions or who were just very desperate to stay alive.

Mufflan believed that establishing certain types of pavilions would be an effective means for reducing the impact of crime on society. The pavilions were intended to be places where people could go to vent their most destructive impulses on others, who also chose to be there for the same purpose, instead of venting their violent impulses against those in society in general. Of course, these pavilions, like many of the others, were a kind of social experiment–as was the entire pavilion system. Of course, there were no guarantees that these pavilions would achieve their desired purposes, but it was already an established fact that law enforcement personnel seemed to lack adequate means to prevent and reduce crime; often the job of such personnel was to sweep up the damage after it had been done and lock up the perpetuators, if they were found, and too frequently they were not found.

Those who wanted to commit rape, beatings, homicides, and many other acts of violence could do so against those who, by going there or being forced to go there, made themselves into potential victims of violence—at the pavilions. Homicide included suicide, which required not only the participation of those who wanted to be killed but others who were willing to do the killing.

Those people who had criminal records and were unable to find decent work in order to earn a living often became so hopeless in finding a way to cope, but now they would now be able to volunteer to staff the criminal pavilions on a subsistence basis, that is, working there in exchange for room, board, and a small stipend.

Mufflan also realized that something had to be done to deal about white-collar and other nonviolent criminals. He believed that current sentencing laws for offenders were way off base and totally inadequate in preventing crime. For instance, if someone committed a violent crime against another person, the punishment was usually very severe. On the other hand, if someone committed a white-collar type of crime that harmed thousands of people, the punishment was usually mild. Mufflan believed that punishment should be based on the amount of harm and damage that criminal acts inflicted on others. Some victims of white-collar criminals would undoubtedly go to these pavilions in order to pay-back those perpetuators, who had been sentenced there as punishment, and take it out of their hides.

The misdeeds of non-violent criminals were hurting an awful lot of people and undermining the confidence of the public in business and governmental institutions. Mufflan felt that he had to find a way to shake up these offenders really severely so that they would be dissuaded, after previously merely getting off with slight slaps on the wrist, from committing additional crimes. Sentencing them to serve or work in the pavilions might be the answer, and, once sentenced there, if they could be truly convinced that, if they committed additional crimes, they would be sentenced to more severe punishment. Incarceration needed to be, except for the most dangerous offenders, a place to rehabilitate petty criminals so they wouldn't commit additional crimes and become hardened criminals—not a place to cause them to become more embittered and hardened so they would want to go out and commit additional crimes. Those few criminals, who were truly dangerous, had to be removed from society permanently by confinement for life or even death sentences; in most other cases, the pavilions might be able to get rid of those types of criminals quickly and permanently.

Mufflan believed that the presence of dangerous criminals would even enhance the quality of the pavilions, where violence itself was the attraction, and where paid visitors would have opportunities to deal with those who had not been law-abiding in society. Of course, these violent offenders had to be carefully watched, because not only would some try to inflict violence on visitors at every opportunity within the pavilions, but also outside of designated activity areas. Some offenders might also try to harm those who maintained and guarded the pavilions facilities. Perhaps a system of rewards and additional punishments could be used to rehabilitate violent criminals or at least make them less dangerous; those might be monetary, sexual, narcotic, or some other type of inducements. If those criminals could be motivated to direct their dangerous impulses only in the designated areas within the pavilions against authorized participants, they might receive rewards for limiting whom and when they harmed others, and also, accept the fact that they—the most dangerous criminals--could never leave the pavilions, the incidence of violent crime in society as well could be reduced significantly.

But in specifically-designated areas of the pavilions, those who wanted to commit violent acts would have ample opportunities to freely shoot, beat, stab, straggle, and kill pavilion visitors by those and other means as long as they did not harm staff members of the pavilion. There was one way to help minimize the risk to non-violent employees at the pavilions; that fact was that some of the criminals themselves would be employees too and would be compensated for what they did, provided that they obeyed the rules. Out in society their acts of violence would have been viewed as criminal, but there in the pavilions, these were what they were supposed to do against those who paid to be there, as well as a number of individuals who would be sent to the pavilions as punishment. But unlike in society, in the pavilions, they would be more likely to be victims of violent crimes themselves. Many of those criminals would die at the hands of pavilion visitors or other violent criminals. So even though these violent people still committed violent acts, they had, in a perverse sort of way, been rehabilitated, and because they were away from society, it was a safer place without them. And their behavior was no longer a threat to those who did not deliberately seek it out by going to the pavilions.

Some of the pavilion visitors came to them expressly with the express purpose of giving those violent criminals, whom they considered as deserving of violent treatment, what they considered as their just desserts; some indeed did just that, but some of those they killed were other pavilion visitors.

When someone would visit a homicide pavilion, he or she–yes, a number of women were expected to visit this pavilion—would pay a fee and be given an opportunity to select various weapons including knives, clubs, chainsaws, axes, and several varieties of firearms—from almost any type of weapon or implement that could kill or cause bodily injury. Ammunition would be extra and dispensed at the rate of one round at a time. Some visitors might not choose to take weapons, many being those who would come to the pavilions to be killed (in other words, to commit suicide--with the help of others) and in the a series of mazes, which would form some of passage ways, where there would be numerous places for assailants to hide, many of these people would be attacked, assaulted, and finally killed. But others, not seeking suicide, who chose not to accept a weapon might want the challenge of taking a weapon away from one of the other visitors and then using it on him or using his own bare hands and body as weapons.

In the Intimacy pavilions, (colloquially referred to as the meat market.) Mufflan foresaw having at least three types: one for both men and women, and one that catered exclusively to each of the sexes. Each of these might be further designated in terms of options such as rape, bondage, prostitution, etc.

Prostitutes would be available in the three main intimate areas.

Mufflan reasoned that rape would take place in some of the sexual pavilions. It was indeed an act of violence—but in the pavilions it would be voluntary–well, most of the time. After all, what they tried to do unto others might be done unto them.

Mufflan envisioned some of the pavilions as places where people with anti-social attitudes could be temporarily removed from society in order to be rehabilitated. That included many people who were not technically criminals but those who might voluntarily go there to improve and reform themselves.

And, of course, those convicted of serious crimes against other people would also be sent to these pavilions.

Those who sought to patronize some of the pavilions—gambling, drugs, sex --

would have to undergo background checks before they were admitted to the pavilions to ensure that potential losses caused by their deaths or injuries would not hurt anyone and that people who were dependent on them would be financially provided for.

There would also be a need for a specially trained and equipped security force to deal with problems in the pavilions. In addition to the need for many types of skilled workers to man the pavilions, a large number of medical and funeral personnel would be needed to evacuate, dispose of, and care for the dead and injured.

The physical distances of transporting visitors to the pavilions were a minor consideration, and the distances between the entrance way and screening areas of the various pavilions and the place where the actual activities took place would vary in distance. It was expected that many of those who would frequent the pavilions would do so secretly, anonymously—without the knowledge of even those closest to them; therefore, providing a means to keep peoples' visits to the pavilions low-key and discrete had to be considered.

Methods envisioned for transporting visitors to and from these sites were in the prototype stages. In fact, the pavilions would be a good place to test out new ideas and innovations. It would also be a good place to do research and development work in other areas, such as testing weapons and medical techniques and equipment.

At the transporting stations for the pavilions, visitors would purchase tickets, which included the cost of round trip transportation and, if they desired and knew precisely which pavilion they wanted to visit, entry costs for those pavilions. Voices over the loud speaker would announce the departure gate numbers and times of the various transports to the pavilions, using alphabetical letters to designate the various pavilions. When one arrived at the desired pavilion, they would go to a reception area, and from there visitors would be directed to the desired pavilions. Or, in the reception area of the pavilions, they could choose to go back and get partial credit which they could use at other pavilions at a later date, if they had changed their mind about entering an actual pavilion.

6

PARTNER IN CRIME

Joseph Mufflan 's upset victory over the incumbent in the general election for president naturally provoked many articles and books on a variety of subjects about the makeup and standards of a society, including the following: One issue being the question, what is law? And Mufflan's writings addressed the issue of not only the legislated kind but also moral and ethical law. Another issue addressed the question, what is crime? Over a period of thousands of years, human beings had had many standards for defining what constituted various crimes, as had been defined in established laws, in standards, ethics, values, and, simply, in terms of what was right and wrong and what should be legal and illegal. And the leading provocation for the exploration of these issues was Mufflan's capture of the White House.

"Mr. President, who will you select to implement the drastic changes you have proposed?" a reporter asked Mufflan in one of the president's early news conferences.

"I concede that what I propose may appear to be drastic, if you care to use that term, to some, even to many," Mufflan replied. "There is indeed a need for drastic changes, and it's time to try to make some of those changes. Let's face it: many of our laws have been dismal failures...."

"Excuse me for interrupting," the reporter interjected. "But how so?"

"That's easy to explain," Mufflan said. "Yes. It's really quite simple. Those laws have not achieved what they were intended to do. Look. I believe as president that I can help reform society. And, as president, it is my duty to try to do so. But people-- I can't change their basic nature. Many of them have an underside full of darker urges, which they may repress most of the time. But at other times they're going to try to unleash those impulses or desires. They're going to try to vent those urges, and, during those times, a lot of people, including you or me, may get hurt, even killed, in the process. If there were places like the pavilions, which I propose, a lot of people are going to be spared the harm that otherwise might be inflicted on them by those who need to get these urges out of their system."

In addition to designating appointees for the traditional cabinet posts, Mufflan proposed additional bills concerning the establishment and operation of the pavilions and the appointment of officers who would construct, manage, and operate them.

The pavilions would be a massive project, and many people would be needed to design, build, operate, and maintain those facilities. Shortly after Mufflan had been sworn in as president, full page advertisements appeared in newspapers and magazines throughout the country, such as the following listing:

HELP WANTED
Engineering Medical   Artists
Construction   Security   Painters
LandscapersMechanics   ....and many other types of skills—please inquire.
Great pay, benefits, and working conditions. Flexible hours.

The advertisements for personnel to build and staff the pavilions resulted in a flood of inquiries and applications.

Mufflan believed that in time that the pavilions would provide employment for thousands of people—making a large dent in the number of the unemployed. The operation of the pavilions could also result in the creation of new businesses related to the pavilions, such as eateries, travel services, and lodging.

The following notice appeared in financial publications, seeking investors to finance the construction and initial operation of the pavilions:
INVESTORS SOUGHT FOR NEW PROGRAM
LARGE MARKET—HIGH RETURNS—LOW RISK—GOVERNMENT INSURED
PRIVATELY OWNED AND REGULATED
LIMITED GOVERNMENT RESTRICTIONS

Insiders knew that this notice pertained to raising capital for the "come-as-you-are" pavilions.

The newly created and appointed assistant under-secretary of pavilions knew Congressmen Phillip Anderson, whom he had once indirectly paid off in order to have a business man arrange some clandestine operations for the government. He asked the Senator to get in touch with the business man, who was now serving time in prison for illegal financial dealings, to work with the administration on the pavilions.

"The Congressman left a discrete message with the business man, who returned the senator's call.

"It's Louie Parnelli. Yeah, the guy in the cooler. Will you get a message to Congressmen Phillip Anderson for me. Yeah, he knows where I am; I'm not going any place yet."

The return call from Parnelli to the Congressman went as follows:

"Hello, this is Congressman Anderson," the caller announced.

"No, kidding," the imprisoned businessman replied sarcastically.

"Yeah. So you think you might be my man?" the Congressman asked.

"No doubt about it," Parnelli answered. "I know vice; that's been my business for more than thirty years. Vice, that's what this pavilions thing is all about, isn't it—no matter what kind of fancy words you use to describe them? I know the business. Even in here I still have my hand in it."

"I'll run this by the President. He might like the idea, you know: it takes one to..." Congressman said.

"...Yeah, to know one," Panelli replied, finishing the phrase. You let the President know that I'm interested. You think he can spring me from this place?"

"He can pull strings, if he wants to. I can still pull a few myself," the Congressman admitted.

"I guess you might say, I'm calling in that favor I did for you; you owe me. It could be you instead of me in here. The difference between you and a lot of others is that I got caught," the man told the Congressman.

"Don't get cocky with me," the Congressman cautioned. "I'll see what I can do."

"You do that. And don't be too slow about it," Parnelli replied.

Weeks passed. And Parnelli focused on other matters, such as other possible ways to get out of prison and to make himself as comfortable in it for as long as he had to be there. He had put the earlier phone conversation largely out of his mind until one day he was summoned to the warden's office.

The guard, who had brought in Parnelli, stood by the wardens desk.

"It's okay, the warden told the guard. I can handle this one. Please wait outside," the warden told the guard.

"Yes, sir, the guard replied and left.

"OK, what did I do this time?" Parnelli asked.

"You really think you're stuff hot stuff, don't you?" the warden scoffed.

"I guess I do have a pretty high opinion of myself—even in a place like this," Parnelli replied.

"I guess someone else thinks so too," the warden remarked.

"What are you talking about?" Parnelli asked.

"You've been paroled. It seems that you have some friends in very high places," the warden explained.

Parnelli just sat there, mouth open.

The warden handed Parnelli a package of clothing and said. "Well. Now you go back to your cell and change your clothes—yes, they've even ordered you a new set of clothes. Collect your possessions and come back here as soon as you can. All right?"

"Anything you say, sir," Parnelli replied in a tone that was a mixture of sarcasm and respect, taking the package from the warden and stepping outside the warden's office.

"Bring the prisoner back here as soon as he's changed," the warden told the guard who had been waiting outside.

The guard nodded.

Later, once again, Parnelli sat in a chair in the warden's office, attired in a suit, a packet containing personal items under his arm.

"Here are your reporting instructions," the warden said, handing some papers to Parnelli. "They've even made hotel reservations for you in Washington. These are your plane tickets. Any questions?"

"No," Parnelli replied. "I wasn't sure they were going to go through with this."

"Well. They have. Now get out of here"

And Parnelli didn't have to be told twice.

7

BUILDING & STAFFING 'EM

Once he arrived in Washington and had gotten settled, Louis Parnelli went to the address listed in the instructions he had received when he was released from prison. He entered the office building, took the elevator to the designated floor, and then proceeded to the suite number given in the instructions.

He entered the suite, walked up to the desk, and told the receptionist, "I'm here to see Samuel Hopkins." The name is Louis Parnelli."

"Do you have an appointment, Mr. Parnelli?" she asked.

"Not exactly. But Mr Hopkins is expecting me," Louis replied.

The receptionist picked up the phone and, after conferring with someone on the other end, told Louis, "Please go right in, the door to the right."

Mr. Hopkins rose from his chair when he saw Louis come in. He held out his hand and said, "I've been expecting you, Mr. Parnelli."

Louis shook the outstretched hand. "Please call me Louie," he told Hopkins

"All right, Louie. Please sit down," Hopkins said. "Is there anything I can get for you?"

"Coffee, if it's not too much trouble," Louis replied.

"No trouble at all," Hopkins said. He picked up the phone and spoke to the receptionist.

Soon the receptionist brought in a tray containing a carafe, two cups and cream and sugar. Hopkins filled the cups and handed one to Louis. "Cream and sugar's there," he told Louis, pointing to the tray.

"No thanks," Louis replied, as he sipped his coffee. "This is fine."

"Shall we get started?" Hopkins asked.

"Why not? Oh, yes. But there is one administrative detail I would like to clear up, if you don't mind," Louie announced.

"Okay," Hopkins replied.

"I'm on parole–from prison," Louie continued.

"I didn't think it was from an amusement park," Hopkins replied. Both men laughed, and that seemed to break the ice. "So what's the problem?"

"I'm supposed to report to my parole officer every two weeks, and because of the nature of this job, I'll probably be traveling a lot in this new job," Louie explained.

"Yes, some traveling will be required," H

Louie

Louie

Louie

"This first compound and these pavilions will serve as a kind of baseline for the whole pavilions project. If it is successful, the development of other compounds will follow," Hopkins explained.

"And who will determine what success is?" Louie asked

"Another very good question. One of the major factors is money: how much we have to spend and how much we rake in. The president, the investors, and the public's reaction will also be important factors in determining if more pavilions are planned and built," Hopkins said.

"OK. As I see it, we'll need to hire a contractor who can prepare that land for constructing the buildings on the property," Louis said.

"Do you know someone who can take care of that?" Hopkins asked.

Louis smiled. "Yes, I'll get in touch with him. He can arrange to have workers brought in to get the land ready," Louis replied.

"Good," Hopkins said.

"We'll need someone to draw up the plans for the buildings that will be constructed in the first group of pavilions. I know someone who can take care of that. Of course, I'll need some details on what kind of facilities will be needed to accommodate the activities in each of the proposed pavilions," Louis said.

"I've had the President's ideas translated into some preliminary drawings and instructions," Hopkins told Louis.

"Good. You're way ahead of me," Louis said. "I like that."

"I try to be prepared so I can be on top of every possible situation," Hopkins replied.

"Where were we?" Louis asked himself. "Oh, yes. These properties will need some landscaping, parking for clients, security, and there's some other areas we need to take into consideration."

"Of course. And? " Hopkins asked.

"I'll get in touch with some people to arrange those items," Louis replied. "Once we get some plans drawn up and the land prepared, an architect can lay out those designs in more detail."

"And, I assume that you know someone who can handle that," Hopkins remarked.

"I know a number of good architects, who I've worked with before. They do very fine work," Louis remarked.

"Good, very good," Hopkins murmured.

"I'll get in touch with a general contractor, who can arrange for carpenters, masons, plumbers, electricians, and other skilled workers for this project," Louis said.

"That's great," Hopkins replied.

We mustn't forget the insides of the buildings," Hopkins remarked.

"Of course, not. I know someone who can put us in touch with the best decorators who will arrange and furnish the insides according to our requirements. That can include everything from lighting fixtures, special plumbing needs, beds, couches, anything thing else you can think of." Louis answered.

"Not at this time. Sounds good so far," Hopkins replied.

"And of course, special requirements must be taken into consideration, such as sanitation and ventilation. Can we go back to the actual layout of the property?" Louis asked.

"If you wish," Hopkins agreed.

"It's just that I don't want to leave out essential details," Louis remarked. Hopkins nodded.

"I envision the compound for each set of pavilions being surrounded by chain length fencing, even electrified for greater security," Louis began. "Inside the perimeter there should be plenty of parking for cars. The perimeter and each of the buildings should be monitored by closed circuit television. There should be stations for security guards and also roving foot and motorized patrols. At the entrance of each building there should be other security measures. And there should be personnel within each building who can be readily available to respond to any problems. And, of course, there should be medical and around-the-clock maintenance personnel to take care of any emergencies."

"Anything else for now?" Hopkins asked.

"I'll get back to you," Louis replied. "Is there a place where I can work and make phone calls?"

"Yes, please follow me," Hopkins said. Louis was shown into an empty office with a phone, a computer, and printer on the desk.

"You can work in here," Hopkins told Louis. "If there's anything you need, refer the matter to the receptionist. If it's something major, just pick up the phone and let me know. Anything else?"

"Not for now. I think I already have enough on my plate for now. It may take me a couple of days to go over what we discussed."

"I'll let you get to it then," Hopkins said and left Louis alone in his new office.

As the initial group of pavilions neared completion, Louis Parnelli gave the go-ahead to the decorators, who would furnish and equipment the various pavilions. That included wall coverings, draperies, and furniture, including couches, chairs, lamps, beds, etc.

Mufflan wanted to make each and every person's visit to the pavilions an unforgettable experience. And to help achieve this, he sought to recruit the very best people available to staff the pavilions, while keeping the admission fees relatively low.

Just as had been done in designing and constructing the pavilions, advertisements were placed in newspapers and periodicals throughout the country to recruit the personnel who would operate the pavilions. The following is one of these advertisements:

HELP WANTED

Trained Medical Personnel Plumbers   Lawyers and Legal Assistants   Janitors & Custodial Workers   Managers Foremen   Carpenters   Electricians   Clerical & Receptionists Personnel   Security Specialists (especially former military and law enforcement)   Massage Therapists—Therapeutic and Special Requests   Personal Attention Hosts & Hostesses

....and many other types of skills—please inquire.

Great pay, benefits, and working conditions. Flexible hours.

Message therapist and personal attention hosts and hostesses, to many, meant sex workers. Applicants for these positions were required to submit very detailed

45

applications, which required applicants to submit their sexual history and an elaborate on what intimate techniques they had participated in and were willing to participate in. During follow interviews they would be required to demonstrate their proficiency with various partners. That was done in order to screen out the amateurs from the professionals.

8

TURKISH DELIGHT

Frank Giddings, a reporter for the *The Periscope*, had mixed feelings about his new assignment--at first. God knows, it sounded exciting; it fact, he found himself actually getting aroused just thinking about it. But such an assignment as this one could also be very dangerous, even fatal. And, in the course of his work, to get deep enough into these story, he might have to do quite distasteful things he wasn't sure about doing. He recalled the initial meeting with his boss about this assignment.

"Have I got an opportunity for you," his editor, Earl Murphy, and boss told him.

"You mean a job?" Frank asked.

"Sure it's a job, but what a job. It's not like any other I've ever assigned anyone to," Murphy explained.

"Well, I'm listening," Frank said.

"Say, you're not married, are you?" Murphy asked, seeming to change the subject.

"No, but what has that got to do with this?" Frank asked.

"Any friends, girlfriends, boy..., what I mean," Murphy stammered.

Frank shook his head.

"You don't like...?" Murphy continued.

"Sir, if you're insinuating that I may be..." Frank stammered.

"Gay, you mean?" Murphy asked, finishing his explanation.

"I haven't really thought about that very much. I've done it with girls, even guys a few times. I didn't seek them out; it just happened. I enjoyed it at that time, but I was worried about losing my job or getting into trouble, if anyone found out about those times. I was in the Navy for a short time. There were long periods at sea with all male crews. Well, you know, things happened. But that doesn't mean you're gay," Frank explained.

"It's okay. Your secret is safe with me. I just knew you were the right person for this assignment. So you'll have a chance to do some exploring. I want someone for this assignment who is open-minded," Murphy explained.

"What are you talking about?" Frank asked.

You've heard about the pavilions, you know, the come-as-you are places?" Murphy asked.

"Who hasn't," Frank answered.

"Yeah; that right. There's been a lot about them in the media. But what do you really know about what goes on in those places?" Murphy asked.

"People do talk. And I can imagine..." Frank replied.

"So can anyone. And I want you to put yourself into a curious frame of mind and find out personally. Yeah, let your imagination run wild, but also keep a discerning perspective on what you observe" Murphy told him.

"I still don't see what you're driving at," Frank said.

"I've heard that a lot of sex takes place in the pavilions. There used to be, I guess there still are, places like that, but those were only for men," Murphy told him.

"What kind of places?" h Frank asked.

"Those kind of places have been around for years. They're called Turkish baths, baths, bathhouses. Exclusively for men. Get my drift?" Murphy asked.

Yeah. I've heard of them; never been to one though," Frank said.

"Well. Now's your chance. One of the pavilions is supposed to be like the bathhouses. I suspect that the ones still out there in society will be closing soon—if the pavilions are successful-- because they won't be able to compete with the lower prices at the pavilions. I need to get someone in there who can do a story straight from the source, from a personal, a close, and I mean a real close, point of view. Someone's got to go in there and find out," Murphy explained.

"I can imagine what goes on in there," Frank replied, feeling himself getting aroused by the thoughts that were running through his head.

"Sure. People can imagine anything," But you won't have to imagine once you get inside. I want you to do some research at those places, and I don't care what you have to do or who you have to do it with to get the story," Murphy said, smiling, and looking intently at his reporter.

"So you're assigning me to go inside and get the story? That's what you mean, isn't it?" Frank asked. Then there was silence for a few seconds.

"Look at the opportunities, professional and personal," Murphy promised. "There isn't any law that says that you can't enjoy yourself while you do your job."

"Okay. I'll do my best," Frank said.

"I realize that a lot of reporters would turn up their noses at doing this story," Murphy conceded.

Many would, Frank thought; yeah, and how many could really go through with it. To get the story, he would have to become part of the story. A lot of reporters just couldn't cross over that line. But Frank could and even looked forward to crossing it.

His assignment was to visit the come-as-you are pavilion know as Turkish Delight, a type of place he had never ventured into before. He realized that similar places had been around for a long time, but since the enactment of new laws allowing once forbidden activities to thrive legally in specific places under prescribed conditions, and for a fee—a kind of sin tax, many people thought--a variety of pavilions had proliferated in various locations.

Until now, Frank's sex life had been celibate most of the time. He masturbated occasionally. And, while on foreign assignments, he indulged in intimate pursuits that were cheap and readily available. And there were times when he worked out at the gym and would meet someone, and one thing would lead to another. He knew he might have to open up a lot more to get the full story for his magazine at the place he was being sent and even other such places after that.

While the taxi transported him to the pavilions, he thought about some of the other possible assignments in other pavilions. Some of them could be outright dangerous—sex, murder, brutality, drugs—who knows. One could go in one of those places and just disappear without anyone ever finding a trace of their body, or so he imagined. Could that really happen, he wondered; or maybe he was just exaggerating.

When he arrived at one of the Intimacy Pavilions, he was given a brochure with the available options from which he had to choose.

He pointed at the Turkish Delight option. He was asked certain questions and signed a form, which included disclaimers to limit the liability of the pavilions and the government. In his case, the form made it very clear that he was about to enter a pavilion willingly; he had heard that many others, who went to them, had not gone there willingly. Other pavilion entrants were not volunteers but were sentenced to them as punishment.

The person at the desk called in a young man and said, "Please escort our guest to the TD suite."

"This way, sir," the attendant told him."

He followed the attendant through a door and down a long hallway, with doors on both sides. The attendant punched in a code at one of the doors, which then seemed to open on its own. "Please go inside where you will be served," the attendant told him.

He nodded and entered, and the door closed behind him.

He walked up to an enclosed cage with a person sitting at a raised desk inside. "A friend told me about this place, a very good friend. He said it was a very good place to meet people," Frank told the clerk.

"Indeed it is. You'll meet all kinds of guys, 'er people, who will be very glad to meet you," the clerk replied, winking at him.

"Sounds like just the place I'm looking for," he said

"You want a temporary membership or one for six months?" the clerk asked.

"Better make it temporary. I'm not sure about my work schedule," he replied.

"I understand," the clerk said, as he filled out the forms. "Would you like a room or a locker?"

"I'm on a tight budget. Which costs less?" he asked, thinking of the limited expense account he was on.

"A locker is less," the clerk replied.

"That would be fine," he said.

Frank paid the attendant, who then slipped a towel, locker key, and a condom through the cage's small opening.

"What do I do with...?" he stammered.

The attendant smiled and said. "First time, huh? Undress and place your clothes in the locker. Wrap the towel around you...and then do....whatever."

When the buzzer sounded and a light flashed at another door, Frank opened that door and walked into the locker area. A man a few lockers away from his watched him.

"Not bad," the man said, as he watched Frank remove his clothing and place it in the locker.

He blushed and quickly wrapped a towel around his waist. "Oh, thanks," Frank said, smiling back. "Hey, this is my first time here. Could you show...?"

"Sure. Be glad to show you the ropes. Follow me," the man told him.

They walked through the dimly lit hallway, on both sides of which were cubicles, some of the doors closed and others ajar, the sound of moaning and movement coming from them.

"What are those?" Frank asked.

"Rooms...for those who want a little more privacy," his host explained. "You'll see that not everyone cares about that. They do it anywhere, don't care who is watching."

Through some of the open doors, Frank could see men lying or sitting on small beds in various positions and state of undress, some appearing indifferent, others urging those outside to come inside and join them.

One man went inside a room and closed the door behind him. Soon another man, without knocking, opened the door, went inside, and closed it behind himself. He heard a loud click.

Frank observed one man glancing into various rooms, moving from one to another open door, and looking inside. One room contained a naked man who lay on his stomach on the bed. The man stopped outside the cubicle, looked inside, went inside and closed the door behind him. Again he heard a click as the door locked.

"Where I'm taking you is called the orgy room," the host explained.

"The org---?" he stammered in reply.

"Exactly. You heard me right. And it's totally anonymous and quick," the host explained.

"Really?" he asked.

"Yeah. And some folks like it that way," the host said. "Everything, and I mean everything goes on in there. I guarantee you: everyone who goes in there hard and full will come back out soft and totally drained dry."

He started to walk away and then turned back.

"It's there for those, who like that kind of stuff, for those who choose to participate. And spectators are welcome. Some like to do it while others watch. Come on; don't worry. No one cares who sees what. If they did, they wouldn't do those things here. It's so dark no one sees much of anything that goes on—at least, not whose doing it or getting it done to."

They entered the pitch dark area and walked to the other side of the room. They stood there, allowing their eyes to adjust to the darkness. Before his eyes had adjusted to the darkness, he heard moans, groans, and whispering.

In the corner two men were kissing and fondling each other's naked bodies. In another area a man was bent over the crouch of another man, who lay on his back.

Another man was lying on his stomach, another positioned over him.

A half dozen or so men were spread out on some mattresses on the floor to form a kind of circular daisy chain.

Someone entered the area, his locker key jangling. The guide jangled his key in response.

The man walked slowly over to the guide and, not saying a word, inserted a hand under the other's towel. The two men walked away together. The reporter watched them.

The guide loosened his towel and let it drop to the floor. The other man kissed him on the mouth, and his head quickly moved and kissed down his body, and when it had descended below his waist, began to move up and down, slowly at first and then faster. Finally the guide let out a cry of release, the other man continuing to move his head up and down for some time.

The man wiped his mouth with the back of his hand and stood up. The host began to reciprocate, kissing and then moving his head down the other man's torso until he was kneeling in front of the other man. The reporter hadn't noticed that someone else had entered the room. His shaking hand had caused his locker keys to jangle, and the man came over and put a hand on his shoulder.

"No," Frank, the reporter exclaimed at the advance.

Frank rushed out of the area, intending to leave before he did something that he would regret. The sign "STEAMROOM" beckoned him to go in. He entered and, once he had found the darkest part, he lay down, a towel still wrapped around his waist. Soon a shaft of light cut the darkness as someone entered the room. He lay there quietly and waited.

Frank felt a hand on his leg, and then it began to massage his thighs, and when he did not express any objections, a hand pulled the towel away. He felt an unshaven face on his abdomen and then moaned in response to the sudden and hard engulfing of himself in by another person, his stomach and legs contorting, his breathing speeding up as he neared the point of release and then emitted a loud cry. When the man stood up, Frank recognized who it was–his guide.

"You?" the reporter cried out.

The guide nodded and laughed. "You don't have to reciprocate. It's all in a day's work."

"Work?" he asked.

"Yeah," the guide said. "I get paid to work here—to suck cock or take in up the wherever."

Frank fumbled for answers and even questions and found none.

The guide continued his explanation. "They call it the pavilions; I call it paradise. If it wasn't for Mufflan, I'd probably be in jail or dead. You mean you really didn't know?"

52

"Know what?" the reporter asked.

"My job is to help liven up this place," the guide explained. "Everyone who comes here isn't gay or even bi. Some guys just want to find out what it's like; so I help them do that."

"I still don't see..." Frank said.

"Let me give it to you straight: The pavilions hire professional, experienced people for the intimacy pavilions, and I'm one of them," the guide explained.

"Oh. I see," Frank replied.

"Do you really. I am a prior sex offender," the guide explained. "I'd been jailed before and was in prison when I got hired."

"You mean pedophilia?" Frank asked.

"No. All my partners were of legal age," the guide replied. "Unfortunately I got caught a number of times doing it out in public."

"So how did you find out about this kind of work in the pavilions?" Frank asked.

"We do get newspapers and magazines in prison. I saw an ad for the pavilions; I applied, and here I am," the guide replied.

"I didn't realize that the pavilions were going to use paid 'er," the reporter admitted.

"Prostitutes, whores; spit it out. But yes. Now you know that they do," the guide admitted. "I got to keep moving; there are other visitors who may need me. We're not supposed to spend too much time with a guest."

"Yeah, thanks for talking with me and doing what you did," the reporter stammered. "I'm not really into this, but it felt good."

"No sweat. Glad to hear it. Come again," the guide replied, giving him a quick wink of the eye.

Then the door opened and closed, leaving the reporter alone in the steamy darkness.

Frank lay there for a few minutes, then got up, showered, and dressed. He had found out enough of what he needed to know, in fact, even more, perhaps more than his prospective readers would have liked to know.

It had felt good, but he would not go back again, and didn't have to. He had enough material to write about at least one of the pavilions. At the hotel, in the shower, he scrubbed himself clean until his flesh felt raw.

When he wrote up his story, he could leave himself out of the activities that he observed and even participated in or at least tone them down or gloss over them, and show what went on there from strictly an observer's prospective. There were still other pavilions to visit, how many he wasn't sure.

9

AH, THERE'S THE RUB!

"Tell me about some of the other pavilions you visited so far," Earl Murphy. The editor, told his reporter Frank Gidding, who was also expected to make some kind of more formal report on each of the pavilions he visited; these would eventually appear as one or more periodical articles.

"I went to the massage pavilion," Frank replied.

"Yeah, and how was it?" Murphy asked.

"Well. The administrative procedures are similar to other pavilions," Frank explained.

At the window of the Rubdown Pavilion, the clerk had asked the reporter which option he wished to avail himself of, the choices being: massage by a female for either a man or woman; massage by a man for either a man or woman. He chose the by-a-woman for a man option.

"I guess clarifying and grouping the options makes the paperwork easier," Murphy remarked, "and also helps customers to decide what they want."

"I guess so," Frank agreed. "There were a number of variations to choose from, in terms of the sex of the massager. I chose to be massaged by a woman,"

"Okay," Murphy said.

"There were further options to choose from" Frank continued.

"Oh, yeah?" Murphy remarked. "Like what?"

At the pavilion he had been asked whether he wanted the therapeutic or erotic type of massage. He chose erotic.

"I choose the erotic type," Frank admitted.

"Good choice," Murphy replied. "Then what?"

At the window, he had been given a towel and a plastic bag in which to put his valuables and was instructed to go to a room number.

He had entered the door with the number he had been given and at the window showed his pass. He had been directed to a cubicle, where he undressed and placed his valuables in the plastic bag that had been given to him. After he took a shower, he lay on his stomach on the massage table, a towel draped over his more private areas.

Frank recalled that he had dozed off when a female voice had awakened him. "I'm Susan, and I'll be your masseuse," she had told him.

"Well, the masseuse came in and started the massage," Frank told Murphy.

"You won't need this" she had told him, removing the towel which had covered the lower part of his body. "I like to give a thorough and complete rubdown," she said.

"Thanks. I appreciate that," Frank had replied.

"What else do you appreciate?" she had asked, reaching between his legs and squeezing his more sensitive areas. He had moaned in surprise.

The masseuse had proceeded to massage his neck, face, shoulders, back, arms, feet, legs. Her hands began to lightly work on his thighs, then moving up, exploring other areas of his body gently. Her firm but sensitive motions on his spine had caused him to convulse and squirm.

"Please turn over," she had told him, and he had done so.

"What do you like?" she had asked him.

"Oh, I don't know," Frank had muttered.

"OK. Well, you let me worry about that. I think you will like what I will do," she had told him.

Her hands had rubbed his chest, and she bent over him, and her head slowly moved toward the more sensitive areas of his body.

"Oh. Oh," Frank had moaned.

"You like?" she had asked.

"Oh, yes. Please don't stop," Frank had replied.

"So how was the massage?" Murphy asked his reporter.

"Oh, it was very nice, really thorough," Frank told his editor.

"I wish I could go there," Murphy mused.

"What's stopping you?" Frank asked.

"I'm too well known. If I did, I'd catch hell from a lot of people in this business," Murphy explained. "Great job. You're the kind of reporter I need–someone who isn't afraid to stick out their neck, or whatever, to get the story."

"Thanks, boss. I must admit that I've mixed feelings about this assignment," Frank admitted.

"How so?" Murphy inquired.

"I had my misgivings even before I started this assignment. Now I look forward to continuing it; I guess the pluses have outweighed the negatives significantly," Frank admitted.

"So keep doing your research; you know what I mean," Murphy said, winking at him.

"Yes, sir, I will," Frank replied. "Anything else?"

"I'd like you to be thinking about how you want to present these subjects in a series of articles in *The Periscope*," Murphy told him. "Keep that in mind, as you gather more data."

"I will," Frank agreed and then prepared for his next assignment in the pavilions.

--10--
GETTING HIGHER

Frank, the reporter, and Earl Murphy, his editor, were reviewing information on the reporter's visits to the pavilions and discussing the next visit of the reporter to the pavilions, this time to the one where visitors could partake of a variety of drugs.

"I was never really into drugs that much," Frank told his boss."

"That's good," Murphy replied. "If you were, you probably couldn't do a very thorough job of observing and reporting on what goes on in the pavilions. You can't get the story if you're all doped up."

"I tried pot a few times," Frank admitted. "I guess most of us did back then."

"I'm listening," Murphy replied. "Go on."

"During those times–and that wasn't so long ago–my focus was on school and acquiring the tools I thought I needed to get ahead," Frank told his boss.

"Very commendable," Murphy agreed. "And now it's paying off."

"Yes, sir. I guess it is," Frank replied. "I've never been a heavy drinker either; oh, I've been bombed a few times.

"That's good, real good; you're going to need a real clear head for this one and an open mind to do a story about those who indulge in order to reach a very cloudy out-of-it state," Murphy explained.

"Yes, I've heard that some of those people who've gone to extremes were trying to reach a higher consciousness," Frank remarked.

"That's one way of putting it, I guess," Murphy replied. "Some critics have said that those who over indulge are just trying to escape reality."

"I guess it's my job to get the story with as much detail as possible; in other words, to get into that reality surrounding what goes on there," Frank added.

"Precisely, without going too far into the moral implications of such behavior," Murphy replied. "I'm pretty sure that readers of our articles will supply their own moral views of the subject."

Unlike other pavilions, for instance the shooting gallery and homicide one, the drug pavilion had not been designed for social encounters, but for isolated indulgence

At the reception desk, visitors were informed of the various options and given a chance to select a suite that offered the types of substances and modes of delivery that they desired. The basic questions that participants had to answer were: what types of substances they wanted and how they wanted it administered to get them to a certain state

Frank realized, that in order to get the story, he had to minimize his own use of mind-altering substances while at the pavilions and yet give the appearance that he was on something, and that included giving such an impression to officials in the pavilions. So he smuggled in a small flask of strong booze, easily disguised as a bottle of aftershave lotion; a nip here and there would help him to appear turned on by uppers, downers, or both.

It would have been suspicious for him to have been observed taking notes at the pavilion. So, to do this he would from time-to-time adjourn to a stall of a nearby rest room to jot down the essential facts in a small notebook hidden in his toilet kit. Even while a student in general studies and later in journalism, he had learned to quickly and efficiently screen information for essential details. He had learned to listen carefully, rather than simply asking a lot of questions, in order to not come on as too inquisitive to the person he was interviewing. In many instances, this reporter had found it unwise to take notes when talking to a source; many people became nervous and standoffish once a reporter started jotting down their words. In the past when he had begun to record a source's comments, a coldness seemed to suddenly creep over the interview, as if committing speech to writing somehow solidified those words in stone.

In his mind, he carefully noted the types of mind-altering substances that could be obtained and used at the pavilions. For each suite, visitors were presented with a menu from which to make their selections, including: Types of indulgences--marijuana, various grades; cocaine; opium, in a number of forms from morphine to heroin; and how one chose to receive it: smoking, snorting, taking in pill forms, shooting it via a needle. Users could sit in comfortable chairs, at tables, or recline, as they partook in the suite in which they chose to indulge. Manufactured drugs might be introduced into the pavilion later, if there was sufficient interest in these.

He could tolerate the marijuana suite easily, but the others--probably not. So he decided to try to converse with those who went into the suites offering the strongest substances.

59

At the pavilion, the reporter saw a lot of sick people, some throwing up, others simply looking awful, some seeming very hyped up, others seeming to be down in the dumps, and many others just lost in themselves, each reacting individually to whatever substances they had put into their bodies. In many, if not most, cases the drugs had caused them to achieve a state they wished to be in and now they basked in their own private world of sensations and senses.

Medical personnel quickly evacuated those who had over-indulged from the visitors' area of the pavilions, where they could receive the necessary treatment.

Frank tried to remain as nonchalant as he could when he talked to those who had come out of the suites and to others before they had gone in.

Frank had seated himself in a very plush chair and was dozing when he heard a voice. "Hey."

"Excuse me," Frank said. "Did you say something?"

"Just making conversation," a man replied, sitting in an adjoining chair. "I said 'hey,' Have you been here before?" the man asked

"This is my first time," Frank admitted.

"I could tell," the man replied.

"Obviously this isn't your first time," Frank said.

The man shook his head. "Hardly. I've been here quite a few times."

"If it's not getting too personal, can I ask you what you use?" Frank asked.

"Oh, you want to know what my chosen poison is," the man chuckled.

"If you want to put it that way--yes," Frank replied.

"I don't mind talking about it at all," he man replied. Me, I like to smoke a little grass to loosen up. And later, maybe then something that will take me to a higher plane. Maybe, snort a little coke. And you?

"Pot is all I've tried so far," Frank admitted. "But as long as I'm here, who knows. Maybe I'll check out some of the other stuff."

"That's the spirit," the other man replied. "Take advantage of it while it's available."

"Grass, coke--you ever try anything, you know, stronger–I mean really strong?" Frank asked.

"As a matter of fact, I shot up a couple of times," the man admitted.

"Shot up?" Frank asked.

"I forgot how green you were," the man replied. "Shot up, you know, with needles.

But I got kind of scared and stopped it all together. I felt that I was getting hooked. Later, I went back to the milder stuff. You now what?"

"I will when you tell me," Frank smirked.

"Very funny," the man snickered. I really feel sorry for some of these folks–the ones hooked on the really strong stuff."

"I guess it was their choice," Frank surmised.

The guide shook his head. "Yeah, it was at first. But after a while, there is no choice. You know, you can tell who they are?"

"How's that?" Frank inquired.

It's really quite easy. You see it in their bruised arms and other body parts that have been stuck with needles too many times–I mean, they're walking sores–if they can still walk, if they're steady enough for even that. You can see it in the burns from free basing and smoking and almost setting themselves on fire. But, but most of all, you notice their being really out of it, even when they're not high. You get some of that stronger stuff in in your system, and it keeps building up in your body, until..."

"Until what?" Frank inquired.

"They get so sick they no longer have any desire or inclination to do the most basic things; they don't have the energy or willingness to do almost anything. When they get on that downward skid, then it's just a matter of time...," he told Frank. "...to death. It's not a very pleasant end."

"How'd you figure this all out?" Frank asked.

"Oh. I've talked to a lot of those types of people and those who knew them," the man explained. "And, as strange as it seems, it's sad and hopeful at the same time."

"I don't see what's hopeful about that," Frank asked.

"When someone starts using the hard stuff, there's still time if they truly want help in kicking the habit. And with Mufflan's program, they can get help without being thrown into the slammer. And if they want to keep shooting up, they can come here and do it; I mean, it's so cheap," he told Frank. "They can even go out of this life with a little dignity or what in their doped up mind passes for it."

The man continued his explanation. "But enough on that. To me, coming here— it's like a vacation—a chance to get away the usual constrains for a while, for a day, even for just a few hours. So I come here once a month—usually that's all the time I have to spare. I wish I could do it every week or even every day. But then, I wouldn't have something special like this to look forward to. I think the pavilions are going to satisfy a lot of the needs for people with various preferences," the reporter said.

"I see it like that too," Frank agreed.

"I pity those who have turned something that could be a pleasant change of pace into an obsession–I mean drugs; they're hooked, you know, addicted?" the man remarked.

Frank nodded.

"You know, a lot of people would live here if they could, but only a few get the chance. Some are so far gone that they're allowed to stay here as volunteers, not able to do much of anything by then; it's cheaper to care for them here than in a hospital. A few even have paid positions here," the man said.

The man continued. "If it wasn't for this place, they'd be in jail or dead a lot quicker than the dope can finish them off. They should thank their lucky stars for a man like Mufflan." And thanks to him, even the addicts and casual users can still get high and function as well as most people who never use drugs. But to most of us visitors, it's pure escape."

"It's been great talking with you; I think I'll check out a few of those suites," Frank told the man.

The man shook hands with Frank. "For me too. See you around."

Editor Earl Murphy continued to go over the reporter's notes, nodding and muttering as his head moved left to right and up and down on the notepad Frank had given him.

"Pretty good stuff," Murphy finally looked up at Frank and remarked.

"I couldn't get into some of the suites without having to use the really hard stuff," Frank explained.

"I understand," Murphy replied.

"So, you see, much of what I jotted down is second-handed," Frank told Murphy.

Murphy nodded and then remarked, "It's still pretty good. And when you write it up with your style and finesse, it's going to be even better. That one guy you talked to for some time seemed to know his stuff."

"Yeah; his stuff and a lot of other peoples' too," Frank agreed.

Murphy chuckled.

"He's been in those places I couldn't check out personally and tried a lot of that other junk," Frank added.

"Well. Keep up the good work," Murphy said. "I can't wait to hear about your next adventure. Where is it going to be?"

"I'll let you know before I head out," Frank promised.

--11--
THE SHOOTING GALLERY

One of the "come-as-you are" pavilions had acquired the name of the "Wild Drivers" pavilion, and it soon became evident to anyone, who visited it, how it had acquired that name. It was also called the "Shooting Gallery" (SG), and those who went there became not only shooters, who targeted other drivers, but also ran the risk of becoming targets of other shooters, meaning other drivers.

To many visitors, including paid guests, the experiences in the SG pavilion came as startling awakenings. And some of those surprised persons were police officers– supposedly, enforcers of the laws concerning the conduct of motorists, that is, but too often exempting themselves from obeying those very same laws they cited others for violating. At times they took liberties and frequently turned a blind eye to offenders in their own ranks, making themselves a law unto themselves and not subject to the laws they enforced and cited others for disobeying.

There were many people who cynically scoffed at what the cops were doing: outsiders saw these officials as not wanting to do anything that might endanger their own hide or jeopardize receiving their pensions. But that would all change very soon.

President-elect Mufflan had ordered his representatives to meet with law enforcement leaders at both the state and local levels. The leaders were promised additional federal assistance, including financial resources, provided that they would mandate programs to rehabilitate the bad cops among their ranks, including repeated violators of traffic safety laws and standards.

All cars now came equipped with computer chips, which in conjunction with sensors mounted along the streets and highways, could, from central stations, monitor a vehicle's speed and its front and rear distance from other vehicles. Now dangerous drivers could easily be identified and removed from the roads. Law enforcement vehicles were also equipped with the same monitoring chips, but until now most of their supervisors, who could review the driving practices of their officers, had chosen not to enforce the law for officers themselves, that is, not until the appointment of new police chiefs. Scenes, such as the following, repeated themselves in precincts throughout the country.

Officers Johnson and Roberts stood in front of the desk of Bill Samuels, one of the newly-appointed police chiefs. He mumbled as his eyes scanned some papers on his desk and then looked up at the two police officers.

Samuels smiled artificially. "I've been looking over these reports," he said, the smile suddenly disappearing from his face.

"You may sit down," Samuels told the officers, and the two officers did so in the chairs to the left and right of his desk.

Samuels shook his head. "They don't look good, not at all, but the evidence is quite conclusive. They indicate that you guys have been exceeding the speed limits more than 95 percent of the time you were on duty."

"Emergencies, sir," Roberts explained.

"Oh, yeah?" Samuels asked.

Johnson nodded in agreement.

"Emergencies?" Samuels repeated in a questioning tone and then stopped to write some comments on the papers. The two officers smiled in relief.

"And where are the reports documenting these emergencies?" Samuels asked.

"Sir, we're up to our ears in paperwork," Johnson explained.

"You're going to be up to your ears in something else, and I don't think you're going to like the smell of it," Samuels retorted. "The regulations are quite clear: Officers are to document instances when they must violate laws, including exceeded the posted speed limit, to enforce laws, when in the line of duty."

"Okay. We may have overdone it a little bit. But everybody, all of us, does it," Roberts said.

"That's true. Unfortunately, but you guys do it more than the rest of us, way too much. Tell me this: How can you in good conscience cite others for the same violations you've committed, and excuse yourself? Who authorized you to break the law? Can I please hear your explanation?"

Neither officer spoke up.

"That's right, damned right. There isn't a good reason. As of now, we will no longer be doing business as usual. Do I make myself quite clear?"

"Yes, sir," the two men answered in unison.

"Are either of you men aware of how many people die in traffic mishaps on an average day?" Samuels asked.

Both men shook their head.

"I didn't think so. Three hundred and seventy-five. That's right: 375 human beings that we were supposed to be protecting. But how can you protect them when you yourself are violating the law and endangering the very people you're supposed to be protecting?

"And do you know the main causes of those 375 deaths each day?" Samuels persisted.

Neither man spoke up.

Samuels continued, impatient at not getting a reply. "Most of those people died because someone speeded or tailgated, cut in too close, or violated the right-of-way of some law-abiding driver. It's that simple.

"An average of three hundred and seventy-five poor smucks bite the dust every day because someone else takes it into their head to become a law unto themselves. And, you know, that's in spite of the built-in microchips in the cars and the roadside sensors, which warn drivers when they're going too fast or driving too close to other vehicles.

"Three hundred and seventy five lives wasted every day, dead in traffic accidents. God, I hate that word. Traffic accidents, my ass. Accidents? No; inevitabilities is a more accurate word. Or suicides, by using their cars, is what they are. Homicides too because they use their cars to kill others; when they go on their joy rides, they take others with them.

"You know; it's really sad, even pathetic, because those guys have the means in the palm of their hand to avoid killing themselves and others. And do they use them— hell, no, most of them don't, anyway? The writing's on the wall, but their eyes might as well be blind. It's right in front of them on their car's computer screen and in the digitized sounds that warns them," Samuels said, and then his voice assumed a tone of digitized computer speech.

"'Warning! Warning! Continuing at your present rate of speed, you will collide with the vehicle in front of you in 4.25 seconds.' So do they heed the warning and slow down? No. And many of them even speed up, in spite of the warnings, thinking they can defy the very laws of physics like some kind of superman. 'Warning! Warning! Continuing to move forward in your lane at your current rate of speed, your vehicle will be struck by the one to your immediate rear in 2.876 seconds. Recommend that you increase your rate of speed or change into the right or left lane.'"

"How much notice do we get?" Johnson asked.

"Notice? Oh no, I'm not going to fire you guys--not yet, anyway. I could, but you'd probably win in appeal. And frankly, I think there's still hope for you. I sure hope so, for your sake. I'm assigning you two to one of the 'come-as-you-are' pavilions–the wild driver's one. On your assignment, it will be your job to enforce the law there, if you can. Of course, the regular patrons are guys and gals who don't want to obey the law. They've paid for the privilege of seeing how far they can go in stretching the law beyond its limits, even if it kills them or someone else–including you guys."

Johnson gulped deeply.

Samuels smiled. "Yes. Some of those drivers would like nothing better than to off a cop or two, and you know why? Because decent cops try to prevent them from driving like they want to, including violating traffic laws and general safety rules. Remember this: You're cops, and you're expected to lay your hide on the line to protect others, including those who don't want you to protect them or others. You're not out on the street to play it safe. Oh, yes, you play it as safe as you can, but doing your duty must always come first. So you stick your neck out, and you squeak by most of the time, but some time you lose.

"Maybe you guys should look up the word 'peace officer,'" Samuels suggested. "That designation refers to someone who is supposed to keep–not break the law. Your very presence is supposed to serve as a deterrent and motivate others to obey the law, even for those who don't want to. Either way, the pavilions will make you better cops or get rid of you, and that'll also solve the current problems. Any questions? Good. Check with my secretary on the way out. She'll give you the details for reporting to your assignment. And good luck, gentlemen. Dismissed."

When Roberts and Johnson arrived at the designated pavilion, they opened a door with a sign on it; it read "RESTRICTED AREA. AUTHORIZED PERSONNEL ONLY."

"Papers, please," said the uniformed man sitting at the desk. Balding and appearing to be about 50, he examined the forms the two officers had handed to him and finally addressed them once again.

"Chief Samuels called and said you were on your way," he chuckled. "So you're the guinea pigs Sam's going to use to shape up the force. He said that generally you both have good records, but are in need of some remediation. Remediation, well this is just the place to get it. You get good real fast, or you get dead, also fast. Then they scrape what's left of you off the pavement or pry you out of the wreck you were in, and then recycle both you and the vehicle, so your car is ready for the next thrill seekers, who pay for their fun. And they recycle what they can find of your carcass into tomorrow's instant dinner, you know, the ingredient that reads 'miscellaneous meat bi-products;' that's another way these pavilions pay off. Everyone one who comes here, civic employees included, signs a consent form saying that if they are killed or disabled that their remains may be recycled for various uses. Hell, we stopped spending exorbitant sums long ago to keep people alive forever when it'll be as vegetables. I mean, what kind of life is that anyway?"

"What's wrong with him?" Samuels asked about Roberts, who seemed to have turned a little pale.

"I think he's going to be sick," Johnson explained.

"That's okay, Samuels chuckled. "It takes a while to get used to this place. If you survive, you might even get to take my place here. I'm going to retire soon."

"You gentlemen can call me officer "G," which stand for guide. We don't use our own names here. Roberts, you will be officer "SG-1. Johnson, you will be "SG-2." The letters "SG" stand for shooting gallery. It's where you guys will spend most of your duty time. Any questions?"

"What...?" Johnson muttered, unable to phase his question.

"Yes?" Samuels asks

"Nothing, sir," Johnson answered. "I thought I had a question; I guess I didn't, after all."

"Good. Then I'll show you around," G said, getting up from his desk.

G escorted the officers past a number of doors, but opening none of them. Some were labeled as follows: "GARAGE," "MORGUE," "JUNK HEAP," BODY SHOP," "SHOOTING GALLERY."

The door labeled "SHOOTING GALLERY EVACUATION" suddenly opened, and the three men paused, while two other men carrying a stretcher slowly walked passed them; the stretcher's contents were covered with a bloody sheet.

"It's okay; we're not in any hurry," G told them.

"Neither is he," one of the stretcher bearers snickered, indicating the contents on the stretcher. "Yes, sir. His speeding days are over."

"I think I'm going to be sick," SG-2 muttered.

"Don't worry about it," G replied. "It happens to a lot of guys."

G noticed that SG-1 and SG-2 were still staring at the door labeled "SHOOTING GALLERY EVACUATION," as if they expected other stretchers to be carried out any minute.

"You'll get a real close-up look at what goes on in there tomorrow," G told them.

G opened a door labeled "QUARTERS" and said. "You two better get some rest. You're going to need it."

The two men went inside.

"Homicidal maniacs, suicides, road hogs, speed demons—you'll encounter them all and many other types in the SG," G told them, communicating in the same unemotional tone as he had the previous day. "Yes, they call it the shooting gallery. Pretty simple, you'd think, but the question remains: are you going to be the shooter or the target? Well, that's entirely up to you. If you're smart and very alert, you'll get them before they can get you.

"Those guys, and gals too—yes, there are some pretty crazy women who come here--and they've paid to be in there. And you know what? They too have nothing to lose but their lives, but maybe in the process of killing themselves--yeah, it happens quite often—too often—they'll take a few others with them.

"Oh, sure," G continued, "you're probably saying to yourself that you've encountered those types on normal traffic duty, maybe in the accidents you documented. Those guys you've encountered drove the way they did because they thought they could get away with it. And they were usually right. Of course, when they're wrong, they or someone else gets maimed or killed in the process.

"And you two, you know why you're here. You were paid to enforce standards that have been set—many of them laws and some just common sense. But since you couldn't beat 'em--hell, there were too many of them--you joined them. Maybe you didn't mean to, but you did. Hell, let's face it, you'd pull one driver over, and a few yards down the road, hundreds or even thousands of other drivers were doing the same damned thing or similar things as the guy you pulled over. Where's the justice or fairness in that, you ask. Beats me. So here you are, assigned to this wild drivers' pavilion.

"Before I turn you loose in there, I think I should give you some background on this place. One purpose of this and other pavilions is to raise revenue for the government. People will spend money to have fun, even as they try to avoid paying taxes.

"The gentleman who thought up this place is not a libertine, in spite of what a lot of people think about him. Oh, no; he's a reformer--that's for sure--but he knew what he

69

was doing. After these guys come here and get into the action and have had the crap completely scared out of 'em in the shooting gallery--and after they leave--the ones who do survive this place, that is--they drive a lot more cautiously—well, most of them, anyway. Oh, yes, some of them will even pay to come back here again, but most will avoid this place like the plague. They also realize that if they mess up on the road and are caught that they may be sentenced to spend time here.

"And you police officers who survive this experience, you'll go back, I hope, and help the other drivers become better and safer drivers. This afternoon I'll show you what you're up against. Any questions. No. Good. I'll see you back here after lunch."

The dining facility was located in the quarters area. SG-1 and SG-2 met G outside the door of the Shooting Gallery after lunch.

"You know what a morgue is; in your line of work, I'm sure you've seen your share of stiffs, accident victims and other victims of non-accidents," G said, having stopped in front of the door labeled 'MORGUE.' "You'll get to see this place plenty; so there's no need to show it to you now."

G led them to the door labeled "BODY SHOP," and motioned for them to go in. Above the din of the many activities that were going on inside simultaneously, he had to almost yell to make himself heard.

"The auto industry would like to outlaw what we do in here, if it could," G explained, as the two men watched damaged motor vehicle being repaired. "Oh, no. They want car repairs to be as expensive and inconvenient as possible; so they pick our pockets whenever they can. We couldn't get away with it out there, but in here, we have to make repairs quickly and get the cars back on the road. You can see how easily, efficiently, and inexpensively damaged vehicles can be repaired and put back into service in practically no time at all. We charge our visitors less than what it would cost to have those repairs made on the outside, that is, as part of the future admission price they'll have to pay to get back into this pavilion, and we get away with it because the damage seems a lot worse than it actually is. In fact, some of the drivers think they're getting a bargain. I guess they don't mind paying for their fun. Yep, some of 'em call it fun—and, I guess it is, in a very perverse and destructive way. Those who want to play, pay, and that supports a lot of worthy causes."

G failed to mention that a part of the admission price paid for a multi-purpose insurance policy that not only covered death and injury, but vehicle damage as well; each admission to this pavilion included a prorated charge of approximately one week of car insurance.

The RR Stations for the Shooting gallery were surrounding by high fencing and were heavily guarded. Inside the main building, Frank Gidding, the reporter, felt pretty shook up after having been out on the track of the pavilion for several hours; fortunately, his vehicle had only sustained minor damage and had been scratched in a couple of places.

While resting at the station, an official reminded him that he could continue with the same vehicle or obtain another.

In reply to that offer, Frank grinned and shook his head, indicating that he had had enough, and told the attendant that he had to leave to keep an appointment.

The attendant also reminded Frank that this pavilion used a points system to access how much customers would have to pay for their next admission to this pavilion. This addition was based on the damage they had inflicted on their own or other vehicles while on the pavilion's driving track. Drivers were penalized in points for the damage they caused and once they exceeded a set limit, they were restricted from visiting the Shooting Gallery penalty for a minimum of 24 hours or even longer.

The attendant also informed Frank that his additional charge on the next admission would be very small.

"That's nice," Frank replied, having already decided that he would not return to the Shooting Gallery pavilion. After all, he had gone there to gather material for a story for his magazine, not because he enjoyed joy rides.

G led them his charges out the door, and they continued their tour, stopping at a set of double doors labeled "SHOOTING GALLERY OBSERVATION."

"This is where the action is, or perhaps I should say, where it's going to be," he said, opening the door and motioning the two men to enter.

From an elevated platform, protected by a window of reinforced shatter-proof glass, they could see the vehicles speeding by below, the sound of engines punctuated by squeaking brakes and occasional crashes and screams.

"That's where we evacuate the dead and injured," G explained, pointing to a freight elevator doors below. "You may recall some attendants carrying someone out earlier."

SG-1 and SG-2 nodded.

"Those stairs lead to our garage where your vehicles are stored. Tomorrow you will be taken out on the track that you see below. Experienced officers will be with you for the first few days. Later, assuming that you're still alive and fit, you'll go out there on your own."

The next day SG-1 and SG-2 found themselves out in the shooting gallery, their guides taking turns driving in the front seats, and they in the back ones.

"Can I ask a question," SG-1 asked.

"Sure. Ask away," one of the guides replied.

"Are you guys authorized to issue traffic citations?" SG-1 asked.

Both of the guides began to laugh. Then one guide answered. "Oh, yes, we are authorized to do that and much more," one guide replied.

"But, it's not a common practice," the other guide said. "You see; if you get out of your vehicle to write someone up, a vehicle may crash into you–accidentally or not so accidently."

SG-1 gulped.

"If you can catch 'em, and they're still breathing and conscious, you may write 'em up. It's a point of pride to those guys to get cited," the other guide explained.

"But why would anyone...?" SG-1 muttered.

"Oh, they just love to brag about all of the times they got away with something and about the few times they got caught. Oh, excuse me," the first guide said.

Their patrol car pulled off to the side of the road. The co-driver picked up the radio's microphone. "Yeah, this is unit 12. I'm going to need an ambulance and two tow trucks in sector X3."

What they had observed were two banged up vehicles, one apparently having crashed into another and then into the wall; there were apparently no survivors. One of the bodies apparently had been thrown out of a vehicle by the impact, and another was impaled on a metal pole by the side of the road.

"Those guys, who pull or pry 'em out of the vehicles or whatever, make real good money," the second guide remarked. "I guess some of that helps pay their cleaning bills. By the end of their shift, their outfits are smeared with blood and ever which thing that results from vehicles and bodies being torn to pieces."

"They can have their money. That's one job I don't want," the first guide said.

"Those medics have to even pull them off those poles. And would you believe that some of them are still breathing even after being run through or thrown against a concrete barrier."

"I'll put all of these details in my reports," the second guide remarked. We need to take plenty of photos of the scene."

"This is really gruesome," SG-2 remarked.

"No more than usual; it's all in a day's work. We've got other fish to fry," the first guide replied.

"So you see what I meant? Got to catch them first before you can issue tickets—that is, if they're still breathing" the second guide commented.

The first guide continued to listen to the radio transmission. "Ten-four," he said. signing out of the transmission.

"I better take those photos," the second guide said. He got out of their vehicle, quickly snapped photographs from different positions and angles, and got back in the vehicle. "That should do it," he said, as he put the camera away.

"You remember Henry Ford?" the first guide asked.

"I've read about him," SG-1 answered.

"Well, Ford thought that everyone should own a car, preferably a Ford, and have the freedom to go where he wanted," the second guide explained. "So we have the driving privilege—not the right. You guys back there listening?" They nodded.

"Don't you go to sleep on me," the second guide told the two men. "You see, the problem isn't where someone wants to go; that's their business. The problem is how do they get there. And, about that second problem, that is, getting there; we do care about that—very, very much.

"Here, it's even worse because for these guys behind the wheel, there is no destination, nowhere to go, just keep going as long as you can 'til you're stopped by one of us, killed or maimed, or you finally leave the pavilion. Hell, even those lemmings that follow each other over a cliff have some place to go."

"And many come back here," SG-1 remarked. "Why do they do that?

"For more of the same," the second guide answered. "I guess they like it here."

About a week later, SG-1 and SG-2 found themselves in the front seats of the patrol car with the guides riding in the back, observing, instructing, and providing feedback. And then two weeks later, the two officers were on their own. Their primary concern was wondering if they would ever get out of that place—alive. Oh, everybody got out all right. But they both wondered if there was a chance, even a slim one, that they would get out in one piece.

The first day on their own, our officers in rehabilitation did manage to stop and issue citations to several drivers. They also investigated a number of crashes, and called in others to clean up the messes, carry the dead and injured away, and tow away the wreckage.

A car speeded past. The two officers gave chase.

"Maybe we can catch this one in time—before he kills himself or someone else. Let's go," SG-2 blurted out.

The vehicle they were pursuing continued at a high rate of speed, forcing several other vehicles off of the road.

"Use the machine guns; stop them," SG-2 shouted.

SG-1 glanced into the sights in order to guide the fire of their vehicle's front machine guns towards the speeding vehicle. His hand griping the joy stick, he squeezed the trigger, the bullets striking the front and back tires on the driver's side. That vehicle skidded to the side of the road and stopped.

"We're gonna take 'em to jail," SG-1 said. SG-2 gave a thumbs up.

"Jail?" SG-2 asked. "Well, maybe for a day or two. He paid for his fun. He'll be restricted from visiting the Shooting Gallery for at least 30 days, but he'll be back. I can guarantee that. This is worse than I thought."

"Worse? SG-1 asked.

"Yeah. I think I understand why these guys come here. And you know what?" SG-1 asked.

When SG-2 didn't answer, SG-1 said, "In a strange way I'm glad I got send here before it was too late."

"What are you talking about?" SG-1 inquired.

"I'm just saying that if I had kept driving like these guys here, I would have become just like them," SG-1 explained.

"What are you saying?" SG-2.

"Some choose drugs, other booze, others different ways to get off. Wreck less driving: it's an addiction."

SG-2 nodded in agreement.

74

After two weeks of independent police work in the shooting gallery, to SG-1 and SG-2 the routine of one day seemed pretty much the same day-after-day: speeding vehicles, chases, crashes into walls and into other vehicles, dead and mangled bodies, and cleanup crews and tow trucks dispersed along the side of the road, policing up the damage.

After about a month of duty in the shooting gallery, SG-1 and SG-2 were called into G's office. They recognized the documents on his desk that they had presented when they came to the Shooting Galley pavilion.

G stamped and initialed the papers and handed them to the men.

"What does 'RH' mean, SG-2 asked.

"That means rehabilitated. It means that you guys have passed. You can go back to your regular jobs. But let me give you some free advice: try to toe the line. The next time you come here—if there is a next time—you may not be so lucky. You see, we like to make it tougher on repeat offenders. Now get out of here, and don't let me see your faces again--unless it's as paying customers," G said, smiling.

--12--
LIFE IS A REAL DRAG!

As Frank left the magazine's office, he informed his editor that he was going on his next pavilion project.

"Which one?" his boss had inquired.

"Another of the intimacy ones," Frank had replied.

The boss snirked. "Well, good whatever," the boss said through his smirk.

Frank felt relieved that his boss had not inquired further. "He probably thought I was going to check out the gals, maybe grab a little nookie," Frank reasoned. But that wasn't it at all. Oh, he was going to check out the gals all right, but in a quite different way and in a very different manner.

Frank was very embarrassed about this one, to say the least, but the boss would find out eventually, when he received the reporter's expense accounts for this pavilion visit. His boss was going to flip out when he reviewed that statement: bills for women's attire and accessories—dresses, wigs, lingerie. Acquiring these items was going to be a real adventure too. He glanced through several fashion magazines and catalogues to familiarize himself with the manner in which women clothed, made themselves up, and how he would present himself it to pass as a woman.

"This visit to this particular pavilion was going to be a real drag—literally," Frank joked to himself–perhaps using humor to help him get through an experience that, if it failed, would be extremely embarrassing and, if it succeeded, wasn't going to make him want to broadcast what had happened to the world. But then, deep down he was going to think, "wasn't I a clever boy reporter to come up with this technique to get a story, and those in the journalistic business will probably think so too, I mean, going to such outlandish lengths to get a really big one." For you see he was going to the female-female intimacy pavilion, and to get into that place he would have to act, dress, and make himself up as a woman---and that included having fake IDs made to give him a female identity.

When he entered the women's apparel store, a clerk asked him if he needed any assistance. He told her that he might in a little while. "I'm here to buy some items for my girlfriend," he told the clerk. "I want to surprise her."

76

"The reporter held up dresses next to himself, resulting in very strange looks from the staff and other customers. "I do this when I buy something like this for my girlfriend," he told another clerk. "In that way, if I compensate for her smaller size, I can see if something will fit her."

He got even stranger looks when he tried on some wigs and held up some lingerie against his own body before he finally made his selection. At the cosmetics counter, as he considered and finally selected some items to purchase, the clerk had to turn away to stifle her laughter, probably thinking that her male customer was "one of those."

The reporter arrived at the intimacy pavilions and proceeded to the Female-Female one, where he soon discovered that its arrangement was very different from the male-male one. He concluded that perhaps women were not as fragrant about how they expressed intimacy as men were. Or perhaps men were more accustomed to dressing, undressing, bathing, and relieving themselves around other men, because environments such as locker rooms, barracks, and places of incarceration were more common places for males than for females. Instead of having an open dressing area in the pavilion, the female-female facilities had private dressing rooms, which could later be used for physically intimate activities for those who linked up in the common areas.

The reporter proceeded to one of the meeting areas where some women were sitting and talking. Yes, females were big on conversation. The reporter remembered how hard it had been for women to shut up so he could get a story; they wanted to keep conversing on about the most superfluous matters—superfluous, so it seemed, at least, for him. They seemed to use, no demand, a lot of conversation to establish intimacy before they proceeded to physical expression. As a reporter, when he was working around women, he just wanted to get the basics so he could writer his story. "Maybe, this experience will teach me some tricks in making it with a gal or two," he surmised; "I mean I'll do what I have to with dames to get to the really important details about them".

The reporter conceded that men didn't seem to mind walking around dressed only in towels, but women were more discrete about their appearance in intimate situations. Here, women wore not only towels around their bodies, but also robes that had been issued at the reception desk; they could stay dressed in those until they took them off in one of the private rooms where intimate physical acts took place.

When Frank had changed into his robe, he kept his undergarments on, removing them would have obviously revealed his gender---so for much of his visit he sat and talked, but mostly listened to other women. And from what he heard, he could deduce what the women did in this pavilion: a lot of touching, kissing, oral stimulation, even using dildos and vibrators. So he used the clues he was able to pick up to determine what he should do intimately when he was with a woman in one of the rooms He made excuses when a number of women hit on him and continued to observe the other women in various degrees of intimacy and undress. But finally he gave in—not totally reluctantly.

When Frank went into a room with one of the women, she quickly removed her robe and lay down on the small bed. When asked why he hesitated to remove his own robe, Frank, the reporter now in drag, explained that he had incurred an injury and did not wish to gross her out. He had shaved extremely close before he had come to the pavilion, and when questioned about why his face was so rough, he replied that he had a skin condition, nothing contagious but it made his skin rougher than it otherwise should have been. He finally told the woman that he would do what he could to please her. So he kissed woman and caressed as much of her naked body as he could reach. She seemed to enjoy his attention. As she lay on her back, he kissed his way down her body, and she moaned with satisfaction. He apologized for not doing more, but within the limits of his deformed appearance, so not upset her. She voiced her approval of what he had done, and he left her lying on the bed.

Earl Murphy, the editor, was working at his desk, planning out the next issue of his magazine, when his inter-com buzzed

"He pushed the button and said," What is it?"

"Boss, the receptionist replied," there is someone at the front desk who wants to see you."

"Well. Tell him I'm very busy; have him make an appointment," Murphy told her.

"It's a women," the receptionist replied.

"Whatever," he scoffed. "I'm very busy."

"She insists that this is really important; she said that she has a ground-breaking story to give you," the receptionist told him.

"Can't she talk to one of the reporters or assistant editors?" Murphy asked.

"She says that she has to give it to you--personally. What's that?" the receptionist asked the women. "Sorry, boss, but she said that if you don't want the story, she'll give it to some other magazine," the receptionist said.

"Tell her to wait," Murphy said, glancing at his watch. He pushed the button of the intercom. All right, send her to my office."

A few minutes later, a woman walked into the editor's office, carrying a large shopping bag, as well as her purse.

"Won't you have a seat?" Murphy told the woman.

She sat down and, without saying a word, pulled several sheets of paper out of her purse.

"I need you to pay these," she told the editor.

"I thought you said you had a story," Murphy remarked.

"Oh, but I do—a really good one," she told him.

Murphy accepted the papers, and, after examining them, said, "This is an expense report."

"Yes, it is. You know Frank Gidding, your reporter?" she asked.

"Why shouldn't I?" Murphy asked.

"He told me to give those papers to you," she explained.

"And what may I ask is he doing with this stuff he bought--for women?" Murphy asked.

"He charged those items; he needed them for an assignment," she told him

"Assignment? For the magazine? For what—as a drag queen?" Murphy scoffed.

"Well, not quite, but they were for him," she explained.

"Has he gone crazy?" Murphy asked.

"Almost," came the reply, which was no longer spoken in the disguised voice of a female as before but in the reporter's natural voice. She began to remove some of the less personal parts of his attire and make up, and lastly, the wig came off

"You've really gone off the deep end this time," Murphy shouted, recognizing his reporter, Frank Gidding. "I'm going to call security and have you thrown out. You should be locked up," his hand reaching for the intercom button

Murphy restrained the editor's hand.

"Boss, I can explain," Frank said.

"This better be good," Murphy insisted.

"But first allow me to change back into my regular clothes," Frank said, standing up, holding a large shopping bag.

The speechless editor pointed at the door to his private rest room. He tried to calm his own heavy breathing and opened a desk door, removed a bottle and a paper cup. He filled the cup half way and gulped down its contents.

Ten minutes later the Frank returned dressed in men's clothes and began to relate why he had shown up dressed in women's clothes.

--13--
MAKING A KILLING

The homicide pavilion was indeed the place where those with very diverse desires and motives converged, where each could indulge his or her violent desires. Three types of people, whom one might encounter in the homicide pavilion, come readily to mind: Prisoners who had been sent there for rehabilitation and punishment; they wore the letter "P" on a black T-shirt; that letter stood for "prisoner." Suicides, in other words, those who wanted to be killed in the pavilion, wore the letters "PS" on their black T-shirt; the letter stood for paid-suicide. Paying customers, who went to the pavilion to simply kill others, wore a black T-shirt with no lettering on it.

Anyone who knew him would probably tell you that Fred was a solid and respectable citizen, someone who was tired of seeing violent offenders hurt others and too often get off lightly or scot-free. Sure some criminals served prison time, but then they were released, and many, if not most, resumed their violent ways in society, were caught again, convicted, and returned to prison, some again and again. Like many such law-biding people, Fred was angry about society having to go to such extraordinary lengths to deal with these violent offenders. It seeming that decent people had to babysit the offenders, who by now should have learned the difference between right and wrong and should have at least tried to become responsible citizens who would do the right thing most of the time.

Legally, Fred could not deal with those delinquent types of people as he would have liked. If he had, he would have been locked up himself—even if he felt that he would have been doing society a big favor. But now, the pavilions offered a way to take some of those violent criminals out of society-permanently. He also saw the pavilions as a way for himself to do his share in reducing the number of criminals, and he planned to take out as many of them as he could—personally and permanently. In his mind, he felt that he would be doing a tremendous public service by eliminating some of the criminals—sort of like taking out the garbage—and, in his eyes, that's what they were: garbage to be disposed of.

Oscar's first run-in with law enforcement and the penal system occurred when he was only in his teens. After he was released from a youth facility, he tried to go straight, but after being frustrated by the lack of employment opportunities, returned to crime to make a living. A pattern was emerging: committing crimes, conviction, imprisonment, which formed a repetitious cycle. And the crimes seemed to get more violent each time: burglary, armed robbery, assault, manslaughter, even murder, which resulted in another manslaughter conviction and sentence. Now he was a convicted violent criminal, who had waited to be sentenced, but this time--not to serve time in prison--but in the homicide pavilion, where he could prey on his own type, as well as the non-violent and passive types who couldn't or wouldn't defend themselves. And for those who came to the pavilion, wishing to die, he was only too happy to help them check out of a world they deemed as not very kind to them. Of course, the biggest challenge he would face was against those who sought to harm others, as he himself had done for so long, but even in prison he had had to deal with some of the same obstacles he would have to face in the pavilions.

A lot of statistical studies have shown that, unlike males, females were less likely to seek a way to end their lives violently. But like many of both sexes, John felt that life had become meaningless. He doubted that he had the courage, resolve, or whatever it takes for one to do away with himself. So he decided to go to the pavilion to die or, more accurately to be killed—by someone who liked to use violence to harm and kill. In the pavilions he hoped to achieve his own death—violently, yes, but hopefully as quick and painless as possible.

The main part of the homicide pavilion was laid out with areas in several configurations. It contained several mazes, and once one went in it, he found it best to continue forward until he reached the end–if he did indeed reach it alive–or was attacked by someone who may have been waiting in the subdued illumination of one of the mazes. Within these meandering paths one might run across the remnants of an attack: blood or even the mutilated remains of one; these would usually be quickly cleaned up and disposed of by a clean-up crew–armed to protect themselves from the pavilion's participants. Those, who were employed to remove the debris, did their very best to remove, if at all possible, all traces of armed confrontation quickly; later, other maintenance personnel would do a more thorough cleaning, even painting and repairing walls, if necessary, to cover up the stains and damage. Maintaining this pavilion could be dangerous, even with the protection of an armed escort; and some maintenance personnel were injured and even killed.

There were a number of wide-open areas supported by pillars on all sides near the distant walls and also near the center; these could be used for cover and concealment while lying in wait for victims. And there were narrower passages with shorter walls in front of the main ones. Clean up crews had to work quickly to tidy up and remove any obstructions in order not to impede traffic through these passages.

Maintenance of the pavilion was made easier because television cameras monitored almost every area of the facility so that crews could be quickly dispatched to where they were needed.

When a paid client decided to visit the pavilion, they checked in at the reception desk, and after the required administrative procedures had been completed, they attended a presentation on the rules governing the place, as well as being notified of what to expect. Then they were sent to one of the RR Stations (Rest & Re-supply Stations), which were located throughout the homicide pavilion. Here they armed themselves, selecting from a variety of weapons, which included the following:

Firearms, including revolvers, rifles, and shot guns, with ammunition issue limited to one round at a time; knives of various sizes; machetes; clubs; spears; hatchets; axes, bows with one arrow per issue; staffs; boomerangs; sling shots; hammers, including the sludge types; garrotes.

Those already participating in the pavilions, who re-entered one of the RR Stations, had to turn in their weapons they had selected earlier and choose another type before they could go back into the main part of the pavilion where the action took place. The RR Stations also contained an EXIT door, leading from the pavilions. Paid customers could bow out if they wanted to. Prisoners, those sentenced to the pavilion, could not; however, some of those convicted of crimes and sentenced to service in the homicide pavilion could earn points, and when and if they accumulated the required number, they could be paroled and released from the pavilion. If they committed crimes after they had been released and were convicted again, they could be sentenced to the pavilions for life—which translated into a virtual death sentence; they would eventually be killed or die from some other cause. There was a proposal to staff the fiercest part of the homicide pavilion with prisoners who were there for life, who had nothing to lose, and whose only motive was to survive for as for as long as they could–regardless of the cost.

As in some of the other pavilions, security personnel had to go in and bring out the dead bodies and the injured. Needless to say, there weren't very many injured to bring out. If a participant was hurt seriously, the chances were that he would soon be finished off by someone else before he could be safely evacuated. Security personnel were supposed to be protected from harm, but at times they themselves became targets when they went into the pavilions to do their jobs. Even though they wore uniforms with luminous markings, mistakes were still made, or their status was simply ignored.

Participants were safe from attack at the RR stations, which were located throughout the pavilions At each of these stations were exit doors, which were unlocked on the hour and half-hour; they would be unlocked for exactly five minutes during those times, and participants could open it and leave the pavilion. Even if someone had been safe in one of these, they could be quickly be targeted as soon as they stepped out of one of these sanctuaries and went back into the action area.

At some RR stations, exit doors were left unlocked 24-hours a day.

But how really safe were those who went into those stations? If no one was around to witness irregularities that might have taken place (as Frank Gidding, the reporter from the magazine, might have been), well, he could also become a sitting duck for others who wanted to eliminate all witness to certain acts of violence. The only really safe place was on the other side of the exit doors leading out of the pavilion, and if someone attempted to use a weapon or tool to harm someone outside the designated killing areas, they were quickly gunned down by security personnel.

From the time that he checked in to the pavilion, although it offered him what he wanted—that is, a quick death from another participant--John had misgivings about being there and for good reasons. Someone had aimed a pistol at him, and he had waited passively to be shot, but someone else threw a knife and took out the would-be killer. Timidly he slowly walked towards the RR-Station, carrying, or more accurately, dragging a club, his selected weapon, at his side although he had no intention of using it.

Another man appeared. "What have we got here? A real wimp," Oscar scoffed. "Can't take it, huh?" he said, "pointing towards the lettering "PS" on John's T-shirt, indicating a Paid-Suicide . "I'll soon give you what you want." He carried a machete, which he raised in the air to position it to strike downward.

"Go ahead," John exclaimed. Please make it quick.

"Glad to," the man replied.

Then Fred appeared. "Not so fast. But I guess that's your style, picking on a helpless suicide," he said.

"What's it to you?" Oscar asked.

"I don't like punks like you," Fred answered, raising his shotgun, which was, of course, loaded with only one shell. Oscar quickly threw the machete, knocking the shotgun from Fred's hand and causing him to slip and fall over some slimy substance smeared on the hard concrete floor. The would-be killer, rushed forward and picked up the fallen machete, intending to finish off the fallen man.

John, whose only intention had been to die, quietly hurried over to the retrieve the shotgun lying on the ground. He picked it up and, as Oscar looked on, discharged the one round into the man, killing him instantly. Then he tore off a portion of his own T-shirt and used it to bind the bleeding wound on Fred's hand. John helped Fred to the RR Station.

"I don't think this place is right for either of us," John remarked.

"Me neither. Why'd you want to kill yourself?" Fred asked.

"I just couldn't take it anymore," John replied.

"Who can? I even feel that way sometimes myself. If you want to kill yourself, there are better ways of doing it outside of this place. So. You still want to do yourself in or have one of these guys here do it for you?" Fred inquired.

John shook his head.

Frank Gidding, the reporter, who had been hiding behind a pillar, had watched the entire scene. He had seen too much violence and killing already. Tucked in the waist band of his trousers was a revolver with its one live round, still unexpended. He had hoped that he would not have to fire It In self-defense or use it against himself to minimize his own suffering, should he find himself in a hopeless situation. And in a place like the pavilions, that wasn't an unreasonable possibility. He had chosen a pistol as a weapon because he considered it to be a less violent alternative to the other brutal and cruel weapons and tools capable of harming or killing.

Frank had seen more death and brutality than he had expected—more than most people would witness in a lifetime: shootings, stabbings, beatings, bodies impaled; men with a firearms shooting at others and hitting, missing, or only wounding others, and then running like hell to avoid retaliation from the men they had tried to kill, and stopped by an arrow or spear impaling their backs against a wall or pillar; men waiting outside RR Stations to ambush anyone who might come out sooner or later, perhaps even a specific target, or just a random one; hand-to-hand combat with men choking or trying to beat the life out of each other; the wounded and injured being quickly finished off by others or slowly tortured with painful blows or thrusts from sharp objects until they finally succumb and lay still; several men or larger groups ganging up on others until they had incapacitated and mutilated them, and then the men turning on each other with a variety of weapons until one or no one walked away; a sledge hammer pulverizing a human skull, scattering portions of flesh and blood, as if a large piece of fruit, in many directions, not sure if the victims they observed lying there were dead or just wounded, and what did it matter after the head received an overpowering blow from a hammer.

Frank had difficulty in removing one particular violent episode from his memory. He had watched as the two men stared at each other, one armed with a chain saw, the other with a pistol. The second fired his pistol and wounded the other man, but the one with the chainsaw bore down on the other, who sadly discovered that his pistol had expended all of its ammunition. His opponent plowed the chain saw into the chest of the other, sending bloody flesh spattering in all directions, and then severed the head, which rolled a short distance after it had separated from the body. He noticed that another man was watching and moved towards the observer. The new man aimed his automatic pistol and fired, and the man fell and landed on the cutting edge of his own chainsaw, emitting an agonizing scream as the still running chainsaw continued to cut through his leg.

"Nice shot," a voice called out.

The man turned and saw another man emerge from the darkness, carrying a sledge hammer over his shoulder.

"Is he dead?" the new man asked, resting the head of the hammer on the ground.

"I'm not sure," the man replied, his hand resting on the grip of his pistol, which was tucked in his belt.

"Looks like he's still alive. Mind if I finish him off?" the new man asked. When the man did not reply, the newly-arrived man switched off the chainsaw and then raised the sledge hammer, gripping it with both hands, and moved it into position over his shoulder and behind his back, as a baseball batter might do to achieve his stance and increase his leverage. He brought the head of the hammer squarely down on the fallen man's head, smashing it as if it were a melon, spattering blood and tissue for several feet.

"Thanks, pal," the new man said, hoisting the sledge hammer over one shoulder and the bloody head of the other dead man over the other and then walking away.

Frank had had enough. "Besides," he concluded, "I think I already have enough material for a story about this place." For his assignment, his job—and what a job—he had gone to this pavilion, observed, noted what he saw, and now could leave in one piece. He cautiously looked around him before proceeding to the nearest RR Station and escape to his more familiar and usually safe world outside the pavilions.

And he had the consolation of knowing that on this very unusual assignment that he had not had to kill or harm another human being or become a victim himself. Most of the other persons who had come here could not say that—that is, those who were still alive and those who, even if scarred or beaten up a bit, could make it out of the pavilion on their own power.

"It was horrible. Just horrible. I couldn't believe the kind of things I saw there," Frank told his boss, Earl Murphy.

"Calm down; it's okay," Murphy replied. He reached into a desk drawer, pulled out a bottle of brandy and two paper cups, and poured some of the contents into them. "Here. Drink this."

Frank accepted the cup and drank down its contents in one gulp and coughed.

"Hey, take it easy," Murphy told his reporter. "I can imagine the kind of things you witnessed in that place."

"Can you really?" Frank asked.

"Yeah. Sure," Murphy answered. "And I don't have to imagine. I've been in this business for over thirty years, and I've dealt with about every kind of brutal thing that one human being can do to another."

"I'm sorry; I'm just a little shook up," Frank explained. "No, not a little—very shook up."

"That's good," Murphy told him.

"Good? And what's so good about all of this?" Frank asked.

"What I meant to say was that you'll put all of those feelings you've experienced into your writing. I can't wait to read your story," Murphy replied.

"It's not going to be easy to write this one," Frank admitted.

"I think you can handle it," Murphy assured him. "I know you can."

"I'll have to; it's my job," Frank replied.

"That's the spirit. Another drink?" Murphy asked.

Frank shook his head.

"You and the magazine will probably win some awards for this story," Murphy mused, topping off his own cup and then replacing the bottle in his desk. "Over the next six months we're going to release your stories. We'll sell millions of copies."

"Is that all you care about?" Frank asked.

"No; of course not," Murphy exclaimed. "There are a lot of things I care about. But this magazine is in business to make money. And it has to make money to stay in business. Anything wrong with that? I mean, you wouldn't object to getting a big raise out of this assignment, would you?"

Frank smiled and shook his head

"You know the kind of violence and brutality you saw in that pavilion goes on all over the world every day," Murphy told him.

"But that doesn't excuse it," Frank shot back.

"Of course, it doesn't," Murphy agreed. "You know what? I thought Mufflan was out of his head when he proposed building the pavilions. Now I see why he did it, He realized that he couldn't stop those kind of things from happening, but he could reduce their impact on society by channeling those perverse and violent impulses into the pavilions--even making money off of them and using it to pay for much needed programs.

"This magazine is going to publish a series of editorials in support of the pavilions as a social good," Murphy told Frank, who just sat there, silently, with his mouth open.

It was inevitable and was eventually going to happen: the pavilions that had been operating were going to lead to offshoots, that is, in the establishment of more specialized ones. Such was the case of the Gladiators Pavilion.

Some critics—not of the idea of the pavilions themselves but of how they were being operated or what they offered--felt that some pavilions were too limited in their choice and scope of activities. Other critics voiced their objections to the Homicide

Pavilion, many feeling that it was too heartless in its violence. After all, where was the sport in killing those who had come there only to be killed without even putting up a fight; those victims were there in order to facilitate their own deaths—in other words, to commit suicide with the help of someone else. Some critics not only voiced their objections but also offered suggestions for changing or expanding the Homicide Pavilion. Those sentiments were the origin of ideas that spawned the Gladiators pavilion.

Some of those critics thought the Homicide Pavilion was fine for those who wanted to kill each other in a variety of ways. "Let that continue," they said. "But let's also have some real competition, in which strong, fit, intelligent, and eager competitors fight each other—to the death or until they both agree to end the fight."

The would-be competitors, instead of paying a fee to enter the Homicide Pavilion, could apply for the gladiator competitions and earn a chance to win large prizes. Those convicted of crimes and sentenced to the Homicide Pavilion could also volunteer to compete as gladiators, but were not forced to do so.

The idea for the Gladiators Pavilion led to another type of event: televising various events within pavilions, but only those that warranted being watched and were deemed to be of entertainment value; these segments were also used for advertising purposes to bring in visitors to the pavilions.

The contests were televised on a pay-for view type arrangements; later, if sponsors could be obtained, they would be broadcasts to the general public.

Here is how one such event was presented:

The announcer said, "Live from the come-as-you-are pavilions: The Gladiators. Some of you viewers have undoubtedly been to the pavilions already. Others of you haven't and probably wonder what goes on in these places. Tonight, live from the pavilions, a number of pairs of contestants will challenge each other to fight to the death or until they mutually agree to stop fighting.

"And without further ado, let us get to the first contest; one contestant is attired in blue, the other in red. The weapons agreed upon are clubs."

The contestants came out. BLUE swung his heavy club at RED, who parried the opponent's weapon. A club could also be used to jab, stab, and to ward off swings of the opponent's weapon, but to strike effectively to the point of lethality, the user needed to obtain some distance from his target so that he could apply a blow with enough force to incapacitate his opponent. While RED was blocking the blow of his opponent, BLUE struck RED on the chin, stunning and knocking him to the ground. As RED lay on the ground stunned, BLUE's club landed squarely and loudly on RED's skull. While an official of the pavilion raised the hand of the champion in BLUE in victory, workers carted away the lifeless body of the opponent in RED.

The contest was followed by some announcements about other activities in the pavilions and other advertisements from sponsors, whose financial contributions were paying for the event, including prizes for the winners and the survivors of the opponents.

The contestants for the second match entered the arena, both armed with axes. The RED contestant swung at his BLUE opponent, who blocked the blow with the handle of his ax and used the other end to strike RED on the side of the head and stun him. Then the blade of BLUE's ax drove squarely into RED's skull. BLUE backed off when his fallen opponent did not move, and his hand was raised by the referee in victory, as the lifeless body of RED was removed.

Some of the live audience cheered, but many booed their disapproval because the contest had been so brief and not much of a match.

The contestants for the third and final match of the evening came out, each armed with machetes. The contestants moved about the arena, trying to size up each other, finding it difficult to get into a position to deliver decisive blows with their weapons, and for several minutes the event seemed to turn into a kind of dance with each opponent trying to position himself to deliver an injurious or fatal blow. Finally, either BLUE got lucky, or RED got careless, and BLUE's machete cut into the arm of RED, who retaliated by making a deep cut in the side of BLUE's midsection. So, RED was not apparently as careless as it had seemed at first. Both bleeding contestants seemed to hesitate for a moment. RED, although he was bleeding badly could see that BLUE's wound was worse than his, even life-threatening; so he avoided his opponent, hoping to lure him in and cause him to make a desperate move that would make him vulnerable. RED apparently saw how desperate his own situation was, and realizing that he had to take out BLUE before he passed out from the loss of too much blood, charged BLUE with his weapon raised in the air. BLUE quickly moved out of the way, as RED passed by him, and then he firmly brought down his weapon on the back of RED's neck and caused him to fall. RED lay on the ground, his hand releasing his weapon; soon he was lying still.

Televising the contests at the Gladiator Pavilion and some of the other activities at the pavilions would bring in a lot of money for the government to use to finance various programs, to pay off the national debt, and very likely reduce taxes. Even many who opposed the very idea of the pavilions--and were repelled by the deliberate violence, brutality, self-indulgence, and lasciviousness—soon joined in support of the proponents of these facilities. Money and personal gain often speak louder than morality and ethics.

--14--
THE MALE-FEMALE PAVILION

Frank Gidding, the reporter, wasn't dreading his visit to the male-female pavilion—well, anyway, not as much as his previous visits to other pavilions. In fact, he hoped that in these uncertain times of insecure, temporary relationships that he would sooner or later meet a woman he could love completely and physically; there was even a chance that he might meet one at this pavilion. Regardless, he had to go through the motions and visit the pavilions, observe, make a report of what went on there, and transform his findings and observations into something others would want to read about as articles for the magazine he worked for.

After going through the receptionist at the intimacy pavilion and being directed to the M-F one, he found himself surrounded by men and women of various ages and appearances; many seemed as confused as he was. Both male and female visitors wore robes, blue for the males and pink for the females. Throughout the pavilion were couches and chairs, arranged so that couples could socialize–prior to more intimate activities. He looked over at her, her being a very attractive woman. She was sitting by herself and glanced back at him and even smiled; a very good sign, he thought. Encouraged, he went over to her and sat down. They talked, sitting very close together and soon realized that they had much in common and liked each other. They both felt that they should find out if they were compatible in a more intimate way.

She took his hand, and they both stood up. They proceeded to one of the private rooms, which contained a small bed, a bathroom with a shower, a small kitchenette with a small pantry and a refrigerator, both well stocked with food and beverages; that was all included in the price of admission to the pavilion.

They continued to talk, and then clothes, that is, their robes came off. It started with the conventional love making, including massaging each other, and proceeding to the more sensitive areas of each other's bodies. Then there was intense licking and caressing with complete abandonment of seemingly every inch of each other's entire body, as if they were trying to satisfy a tremendous hunger that only their physical bodies could satisfy.

This pavilion was designed for one-night stands and brief encounters–not for lasting relationships but nevertheless something longer lasting sometimes materialized. Their lovemaking led each to want more and more--episode upon episode, punctuated with naps, showers, and refreshments from the well-stocked refrigerator and pantry.

He had surmised from bits of conversation he had heard that some of the intimacy pavilions employed both male and female prostitutes. He asked her if she was paid to be there; she was a little taken back by the question. No, she told him; she came here to meet someone–for fun, yes, but, she had hoped, as he had, for something much more. He was very surprised when she asked him if he worked there as a paid prostitute. He shook his head, flattered because he humbly conceded that he didn't think that he was that attractive or experienced sexually.

"You know what I've heard from some of the other girls–from idle conversation that I picked up. That the prostitutes they employ here are trained," she told him.

"Trained?" he asked.

"Yes. Trained," she replied. .Apparently the most attractive people are hired for the pavilions. They have to really like sex. And then they're instructed by other male and female experienced in co-ed classes. The instructors explain and demonstrate every sexual technique you can imagine and some you can't. Then the students practice on each other until they become experts in helping others experience the greatest sexual pleasure possible."

"Wow," he exclaimed. "I don't think I could do that. I mean, I really enjoy sex, but it's much better when you do it with someone you really care for–someone like you."

"I feel the same way," she told him.

With that preliminary out of the way, they broke the wall of intimacy even more and laughed. They both admitted to each other that they were extremely pleased with what they had done with each other and hoped to continue at some time in the future--if they could find opportunities to meet.

They wanted to see each other again, but both had their own priorities and obligations, she a full-time job for a major company, he a promising position with a leading magazine. They exchanged phone numbers and promised to meet again when they could.

Frank wanted to be honest with this woman for he really liked her, and, perhaps, was even falling in love with her. But how could he tell her that their meeting in the pavilions was the result of an assignment that was rapidly turning in to something much more?

After their meeting in the pavilions, the two had a dinner date. Frank told her that he had to go on an overseas assignment for at least several months, maybe even longer. But they could keep in touch by phone and e-mail.

Weeks later , after the article on this particular pavilion had been written and published, Frank felt that he needed to finally level with his boss, Earl Murphy, because he thought his relationship with this woman was becoming quite serious. Although he had not seen her for several months, he believed that he owed it to the woman to tell her how he felt.

"Boss, I need to talk you," Frank announced, as soon as he had entered the editor's office.

Earl Murphy glanced at his wrist watch and replied. "Okay. I got a few minutes."

"You know, I've met a lot of people in these pavilion-assignments," Frank told his boss.

"That shouldn't have come as a surprise," the editor replied.

"Most of those people were--how should I explain it--ships that passed in the night," Frank explained.

"That quite characteristic of those zin this business," Murphy said. "And I imagine especially in places like the pavilions."

"But some of those encounters become something more, if you know what I mean," Frank continued.

"Go on," Murphy told him.

"I even had a date with a woman I met in one of the pavilions, how should I put it, under very intimate circumstances. We seemed to take to each other. I'd like to date her again," Frank told the editor.

"Those kind of things happen sometimes," Murphy replied.

"I like her a lot; it could even get serious soon. I think it probably will, but I haven't seen her for several months," Frank admitted.

"If you like her so much, why shouldn't it get serious?" Murphy asked

"I didn't feel that it would have been right while working on the pavilions-assignments," Frank his boss.

"I see," Murphy replied.

"But now that the pavilion-assignment is about over, I was wondering if...." Frank

94

replied, hesitating.

Murphy was silent.

"Sir, I don't want to do anything that's inappropriate," Frank said.

"I appreciate your concern, but it's okay," Murphy replied.

"You mean to see her?" Frank asked.

"Sure. Why not? Go ahead," Murphy told him.

"Thanks. If things continue to go okay, we might even get married—eventually – if she'll have me," Frank explained.

"And why wouldn't she?" Murphy asked.

"I was on assignment when I met her at the pavilions. At some point I've got to tell her that I met her under false pretenses," Frank said.

"These kind of things happen all of the time. We can't always choose the circumstances," Murphy told Frank.

"Are you speaking from experience?" Frank asked his boss.

Murphy laughed and finally explained. "I suppose I am. In my younger days, I played the field. I went to a lot of sleazy places, including clubs. I was even doing research on those kind of places for a series of articles for a publication when I met my wife-to-be at a place, where she worked as an exotic dancer. We loved each other, but we also had to learn to cope with out pasts. And we got over it. We even still laugh about it when we reminisce about it, leaving out a lot of the erotic details. Hey, I wasn't a saint either."

"Thanks. I'll break it to her gently," Frank told Murphy. "It's good to have such an understanding boss."

"Enjoy it," Murphy replied. "You've earned it."

--15--
CRITICS AND SUPPORTERS

During Mufflan's campaign for the presidency, his election opponent repeatedly pulled out Mufflan's proposal for establishing the pavilions, at every opportunity, like a knife and, brandishing what they knew was a potent weapon; then cast strong objections into stones of opposition against them, belting him with their criticisms. But Mufflan, not content to merely defend his proposals, even embraced parried those negative remarks, and then did his best to explain the benefits that his plan would ultimately reap for citizens and society alike. He deftly countered with his own questions about what the current president intended to do to solve the country's problems—concerns that had hardly been addressed during his term of office. His opponent repeatedly fumbled for satisfactory answers, which he found extremely lacking.

"Drastic conditions require drastic measures," Mufflan repeated over-and-over, and people listened. His opponents had no new proposals to offer; just the same old tired ones, which hadn't worked in the past, and could give no assurance that they would work in the future.

During the campaign, Congressman Peter Stonewall-Jackson Davis had been quoted as saying, "The American people may be danged fools; just look at some of the crap they've put into public office. But I don't think they're fool enough to elect someone like that Mufflan. As president? Certainly not."

In response to Mufflan's contention that much of the behavior that would take place in the proposed pavilions would be in the "victimless crimes" category, Congressman Davis posed the question, "But what about the suffering and negative impact that the behavior of clients of the pavilions will inflict on their spouses, children, and friends? Were they not victims of others' aberrant behavior? No, there were no such things as victimless crimes, and the government should do all that was in its power to enforce laws that help to maintain high moral standards and punish their offenders." Davis believed that as a member of Congress that it was his duty of the government to establish and enforce high standards.

After Mufflan had been sworn in as president, the criticism of the pavilions increased tremendously, and it was better organized than before the election.

At first, Reverend Jacob Jones did not think about Congressman Mufflan's proposal very much; after all, he had a church to pastor and many other duties and obligations than think about all of the hair-brain schemes those in government put forward. And, why should he? After all, members of Congress often introduced absurd, wasteful, and even immoral legislation, and "And thank, God," he sighed, having the satisfaction of knowing that most of these proposals died before they had gone very far in becoming law.

But Mufflan did indeed win the presidency, and only days after being inaugurated, the bill for establishing the pavilions was passed by both houses of Congress, and Rev. Jones seethed at the news. The pavilions became more than an issue that he had to oppose from the pulpit. He concluded that it was his moral duty to find a way to oppose these abominations in action and to do all that was in his power to stop Mufflan from carrying out his immoral ideas. He vowed to use his many contacts throughout the country and the world that he had made over the decades to protest and even overturn the legislation that would establish the pavilions and get rid of Mufflan.

That very Sunday after the inauguration, from his pulpit, after concluding his sermon, Rev. Jones addressed the matter of Mufflan and what he was trying to do. "Well, brothers and sisters, Congress really went and did it this past week. And the president signed the bill. I'm referring to the one dealing with establishing the pavilions, although Sodom and Gomorrah would be a more appropriate name for these dens of the devil."

"Amen. Here. Here..." voices of approval shouted from the congregation.

Jones smiled and continued. "Thank you for those kind words of encouragement. Where was I? Oh, yes. Those who will visit these places, once they are built and open, will not just be simply traveling to a place; no, they will be placing themselves firmly into the very clutches of Satan."

Reverend Jones was not idle. He was true to his word and organized groups to picket the pavilions, but because these facilities were fenced in and had security guards at the gates, the protesters could not reach the buildings themselves. Clients who drove in and out of the pavilion enclosures did so rapidly and hid their faces to avoid detection. The protesters jotted down the numbers of the license plates of the vehicles, hoping to use that information to harass those who went to the pavilions.

Some of the initial opposition to the pavilions continued in other forms other than by religious groups. There was criticism that the pavilions were being built and operated using tax-payer money. Actually, investors had covered those costs, hoping to reap large returns on their investments. Unfortunately, for opponents of certain programs, distortion of the reality often becomes what opponents deemed "higher causes," that is, deliberate distortion of Mufflan's program was their only alternative to the more positive aspects of what he had proposed.

Some of the more extreme critics wanted to have President Mufflan impeached, but his pavilions program was becoming so successful and appealed to a diversity of human urges and proclivities. It also promised to reduce taxes and lead to establishing a number of new government programs. A move such as impeachment did not have the votes to get through Congress----not even enough votes to have the matter brought up.

Other critics focused on preventing Mufflan from winning a second term, but that would not be easy either since the president's popularity and approval ratings was at record highs. But eventually, the failure of political and non-violent means would lead opponents to consider more drastic and even violent means for getting rid of Mufflan and undoing his plans.

Many people, of all ages and economic conditions, supported the pavilions project, almost as many as there were reasons for supporting them, and some of the motives for doing so were well thought out. Support for the pavilions ran the gamut from pure hedonism to curiosity to extreme humility, with numerous gradations in between. Even though many of those same people would probably never visit the pavilions, they seemed to recognize something worthwhile in what the pavilions might achieve.

To hedonists the pavilions were places where they could go to do almost anything they desired from eroticism, to violence, to self-indulgence in the use of mind-altering substances.

To the curious, the pavilions represented something that those types of people would like to try once or maybe occasionally; Mufflan seemed to have recognized that a great number of people had a curious but unfulfilled side.

To the humble, it was not a question of being for or against the pavilions and what they represented. It was an attitude, which could be summed up in this phrase: who am I to judge what is right for other people.

Some supporters even believed in the possibility that something representing much higher causes could be achieved through the pavilions, but certainly not actually in them. And although most of these types of people had no overwhelming desire to go to the pavilions, and many did not even want to know about the details of what went on in these places, they supported many of the possible end results, that is, programs which could benefit many, as well as society in general. Even the very religious recognized that good could ultimately come out of what seemed like very evil practices, and some of the biggest supporters considered many activities at the pavilions as catering to the worst human impulses.

Earl Murphy, editor of *The Periscope* and Frank Gidding's boss, and some of his associates, as well as other journalists appeared on TV in support of the pavilions. Murphy expressed his views in the following editorial, which appeared in a number of publications:

"I have speculated on what went on the in the pavilions, and, I now readily admit that these impressions were viewed through the former prejudices that I had formed over the years. From eye witness accounts related to me, I have been informed about what actually goes on in these places and wish to thank the courageous experiences and unbiased observations of those, who visited the pavilions and even partook in what they had to offer, for setting me straight.

"Still, even though I may not approve of or wish to indulge in the behavior that goes on in these places, I now readily concede that the kind of activities that take place--with or without these places--will continue. But society is better off because more of that questionable behavior will be confined to the pavilions. And President Mufflan's brilliant idea for channeling and directing those activities into the pavilions has resulted in a large amount of income being derived from their operation that will be used for many worthy programs, some even moral ones.

"Isn't that a kick in the pants, for you? Well, it shouldn't be. I am not a very religious person, but I know that even the Bible contains examples of evil and immoral behavior being ultimately transformed into things that are good.

"The pavilions have also reduced prison and jail populations by channeling a lot of what were once considered crimes--such as violence, certain sexual practices, and drug use—into alternative outlets away from society in general. Yes, even prostitution and, what some might label as deviant behavior, have become more acceptable activities when confined to the pavilions. I believe that President Mufflan is to be commended for his brilliant ideas by not only helping to redirect questionable behavior elsewhere, while raising money that will benefit those who approve of such activities and even those who do not. I applaud his efforts and so should the rest of the world."

One economist wrote about the financial benefits of the pavilions, but before discussing this topic, he pointed out some of the ploys used, especially by the very wealthy, to avoid paying less or, even in some cases, no taxes at all. He wrote "Many of those who have been fortunate enough to live in a society that has afforded them tremendous opportunities to amass vast amounts of wealth do all that they can to avoid paying their fair share and, thus, defund that system that has provided those opportunities to them."

And then the author went on to link those tax-avoidance ploys to how the pavilions could partially, at least, compensate for lost government revenues. The government's share of earnings from the pavilions could be used to fund many existing programs, reduce the deficit, lower taxes, and even give rebates to tax payers. He wrote, "How refreshing it is indeed for tax payers to receive money from the government, which usually seems more intent on taking it away from its citizens." In other words, the operation of the pavilions will result in the government having considerably more money to spend than before—and with tax payers being taxed less. The writer also mentioned that the pavilions will provide employment, for ordinary citizens and for ex-convicts who have paid their debt to society and who now need to find honest work to prevent them from returning to a life of crime.

The economist also pointed out the need for re-accessing the priorities of what the government would and should spend money on. He stated quite strongly that a government should not, except in extreme emergencies, spend more than it takes in. He even advocated that government officials, who run up questionable deficits recklessly, should be held accountable and subject to criminal penalties including prison time and should incur an obligation to re-pay the government and tax payers out of their own pockets. After all, common law has viewed it as a crime to buy something without paying for it; so why shouldn't politicians be subject to the same standard?

The economist also pointed out that in a republic there must be a link between the rights that are afforded its citizens to a willingness of its citizens to support its government financially. He suggested that before the pavilions had been established that many of the citizens of the country had lost touch with the connection between what they were paying in taxes to support the government and what they were actually receiving from it; that was perhaps because many of the citizens could not themselves afford to pay for what they expected the government to do for themselves, and therefore, these people should have had more realistic expectations of what their government could and should do. With the additional revenue brought in from the operation of the pavilions, now many people felt that they were receiving a lot more than they had before, and they indeed were because of what visitors to the pavilions were paying there–a part of which went to the government--to compensate for a reduction in the amount of taxes that was needed by the government. In effect, most of the citizens were receiving more from the government than they were contributing because of the money spent by those who visited the pavilions.

Not all the positive feedback on the pavilions came from professional writers. In a number of "Letters to the Editors" columns, ordinary citizens voiced their opinions. One such person wrote, "I am not a writer or a journalist; I am just an ordinary man who is rapidly reaching retirement age. My wife has become disabled and can no longer perform the physical duties of a wife that a man looks forward too. I dearly love my wife, but I miss those tender moments that we used to share in bed. I had read about the pavilions and envied what went on there. Finally my wife breached the subject and insisted that I go there to satisfy those urges that she realized I must have felt and that she could no longer satisfy. I visited some of the pavilions and now go there regularly–without guilt. Because of these places, my marriage is still strong, and my wife and I can continue to show our love for each other in less physical ways."

Numerous other letters were sent to "Letters to the Editor" columns of various publications–from married and single men and women, from straights and gays, from drug addicts and casual users, from the most violent to the most passive people. The common theme of those letters was gratitude that the pavilions provided a place where they could go and do what they wished to do without fear of arrest or harassment or the scorn of society in general.

After the pavilions had been operating for a while, a society of its own began to form inside them and included employees, regular visitors, and those confined inside for punishment and rehabilitation. In the interests of preserving their own way of life, a kind of guerrilla movement begin to form inside some of the pavilions. Even frequent paid visitors could join these groups as honorary members.

The formation of these groups was a reaction to the holier-than-thou organizations that wanted to shut down the pavilions and prosecute those who had established and promoted them; the inside groups wanted to keep the pavilions in operation. Another purpose of these groups was to work for the betterment of employees of the pavilions, including promoting additional benefits and improving the quality of service and experiences in the pavilions for customers. One of their main arguments for maintaining the pavilions was that, in effect, if they were closed, vices and questionable diversions, now confined to the pavilions, would once again proliferate in society, many of them as an underground and a kind of black market for activities that were now more confined to the pavilions.

A provision of the pavilions bill Congress had passed into law was that quarterly reports would be sent and/or presented to Congress. These were to include information on operating expenses, income, deaths, injuries, and other problems and details concerning the pavilions. Some of the findings in one of the early reports were that:

The number of arrests for violent crimes and drug offenses in society had been reduced significantly.

The number of arrests and violations for prostitution and sexual crimes had been reduced.

The number of reported cases of venereal disease was way down.

Many of the preceding items translated into substantial savings for the government and, reduction in taxes needed by the government, as well as improved conditions, greater stability, higher standards of living for families, and improved conditions for communities.

The biggest item on the report to Congress dealt with financial issues and included the following:

The first quarter of operation indicated that government operating expenses were zero (0). (That can't be, some of the critics exclaimed.: "Let them read on to learn why!" President Mufflan and some of his supporters had smirked.) Private capital had covered operating expenses of the pavilions but represented only approximately 5% of the money brought into the pavilions.

Of the revenue brought into the pavilions, after expenses had been deducted, a percentage was allocated for maintenance, capital improvements, and reserves for the pavilions.

The remainder of the money brought in was divided as follows: 50% went to investors and 50% went to the federal government; no government funds were required to pay operating expenses.

Some recommended options for using the government's share of earnings from the pavilions included the following:

Using a portion of the government's share to reduce the federal deficit.

Giving all tax payers a rebate.

Enacting and funding new social programs.

16

HONORS AND CEREMONIES

Even with the tremendous work load of presidential duties, Mufflan endeavored to keep informed about what was happening at pavilions already in operation, as well as ongoing and planned projects--when, not in person, then by phone, photographs, and reports from persons who had observed what was happening at the sites.

Mufflan believed that especially for the first pavilions that it was important for him to generate as much publicity as possible to bring in as much business and to stir up as much interest in the pavilions as possible.

Mufflan was the guest of honor at the ribbon-cutting for the first experimental pavilion. At the he kept his remarks brief and to the point.

"It is with the greatest pleasure that I dedicate this pavilion, the first--I hope--of many more such pavilions. To those who have misgivings about these places, I understand your apprehensions. "

"Like hell, you do," Rev. Jones scoffed, as he watched the ceremony on TV.

Mufflan continued. "But let me reassure you: Much good will eventually come out of what many of you may perceive as just the opposite. I beg your indulgence; please to be patient. And if you do not fully comprehend the means I am using to achieve something worthwhile, please try to appreciate the results I am trying to achieve. I will say more on this subject from time-to-time. Thank you all for coming," he told the crowd, who then warmly applauded his remarks.

There were numerous newspaper and magazine articles on the openings of the pavilions and the ramification of what they represented. President Mufflan found himself answering many questions on the subject; he was even interviewed on several news programs about them.

Presidential duties and responsibilities required Mufflan to travel to many places. He tried to tailor his travel plans so that he could keep abreast of what was happening at the pavilions themselves and at locations where others were being constructed or planned.

Whenever possible he tried to adjust his schedule of appointments and meetings so that he could attend the ground breaking of new pavilions. Whenever possible, he also attended ribbon cutting ceremonies when a new set pavilions was opened.

At some of these ceremonies, there were also news conferences at which Muff took questions; and even pavilion opponents acknowledged some of their benefits.

When Reverend Jones learned that President Mufflan was to be honored at a ceremony for his work on the pavilions and, in spite of the moral and other objections of some of his opponents and critics, also for the positive results of his program, he saw an opportunity to do something more positive. Even the worse critics would concede that Mufflan's program had achieved a lot of good. "But, oh at such a price!" Jones sighed.

And after Jones discovered that that ceremony was to take place in a nearby city, he saw a golden opportunity—yes, pure gold. He knew the city well and even the building where the ceremony would take place. One of his adherents even worked in that very building.

Rev. Jones picked up the phone and immediately called his follower in question and set up a meeting that very afternoon.

That afternoon Fred Wilkins, the follower he had called, knocked on the door of Jones' office in the church.

"Come in," Jones called out.

Wilkins entered the office, closing the door behind him. "You wanted to see me, sir?" Wilkins asked.

"Indeed, I do. Please sit down, my boy," Jones replied enthusiastically. "Can I get you something to drink, coffee, a soda?"

"No, sir, I only have a little time; then I have to get back to work," Wilkins replied. "I told them I had a doctor's appointment."

"In a way, you spoke the truth," Jones said, his words producing a puzzled expression on Wilkins face. "What I meant was that there is a terrible disease that has infected this nation, and I have made it my life's mission to rid the country of it."

"You mean President Mufflan?" Wilkins asked.

"Precisely." Jones replied. "The president is going to be in the area in the near future."

"When?" Wilkens asked.

"In three days. He will be the guest of honor at the Saint Regis Auditorium. You work there, don't you?"

"Yeah, as a janitor. But what has that to do with him?" Wilkins inquired.

"That's where he will be. And you'll have access to the building, won't you?" Jones said.

"Of course. Oh, you have some kind of a plan to deal with Mufflan?" Wilkins asked.

Rev. Jones opened a desk drawer and pulled out a revolver..

Wilkins looked down at it and then at Jones.

"Take it. It's fully loaded. Those six rounds should be more than enough to get the job done," Jones told him.

Wilkins took the weapon and placed it in one of the large pockets of his cargo trousers.

"I can smuggle the pistol in before the event." Wilkins mused. "I'll retrieve it when he's speaking. He will be speaking, won't he?"

"I expect so," Jones replied. News reports said he was going to receive some kind of award. It's customary for the recipient receiving an award to make a short speech."

"So he'll be at the podium?" Wilkins asked.

Jones nodded.

"I'll go in by the side of the stage, where no one can see me. He'll make a good target because his torso won't be blocked by the podium," Wilkins explained.

"I wish there was another way," Jones said. "But someone has to do it."

"It's okay. I'll get the job done," Wilkins replied.

"Let's pray about this," Jones suggested, and the two men kneeled by his desk as he led the two in prayer.

As the applause died down after the award was announced, Mufflan got up from his seat and walked up to the podium; he accepted the plaque, held it up so it could be photographed with him, and then set it down on the podium.

"Thank you very much," Mufflan said, accepting the plaque. I'd like to say....

Several shots rang out in quick succession, and Mufflan put his hand over his chest, as if to keep the life force inside himself from leaving, and dropped to the floor

Secret service agents quickly grabbed and disarmed Wilkins as he turned to escape.

The ambulance arrived within a few minutes, and Mufflan was placed on a stretcher and carried out. Paramedics did what they could, but he died on the way to the hospital.

17

AND.....STILL MORE WORLDS TO COME

In his personal papers, Mufflan had laid out ideas, guidelines, and even instructions for a number of additional pavilions, including their use to solve many of the country's and the world's basic problems. He believed that the pavilions would continue and expand beyond his own lifetime-- activities being changed and added to suit the tastes of a very diverse human population. Mufflan's notebooks contained numerous ideas for continued study and research concerning the places he had envisioned and founded; he even set up a non-profit foundation to continue his work.

In the papers he had left, Mufflan had recorded his ideas for controlling and channeling such practices as the use of tobacco and alcohol. He even hoped that eventually he could limit the use of these harmful products to the pavilions—as he had done with drugs. If he could not do that, he vowed to "tax those products to death" in order to save lives and keep people heathier. He sadly conceded that some people would go to any length in order to indulge in substances that would wreck their health and eventually shorten their lives, as well as those of friends and family members.

Initially the idea for the homicide pavilion had been as a place where paying customers could fight and even kill other people, most who went there having the same violent inclinations. The scope was quickly expanded as a place to punish and rehabilitate violent criminals; that addition would reduce the cost of incarcerating them and paying expensive legal fees and time-consuming delays for appeals for those who had been sentenced to death; in fact, over the term of a single individual's imprisonment, millions of dollars could be realized in savings for each convict, who was instead sent to the homicide pavilion. There were even proposals for sending white collar criminals and others who committed less-violent offenses crimes to the pavilions. This pavilion also remained as an option for those who wanted to be killed. The pavilions, as a place for mercy killing, would also been studied in the future.

The gladiator matches continued as part of the Homicide pavilion; televising them became even more wide spread, and larger audiences watched them live at the pavilions.

Mufflan's proposal for bringing gambling to the pavilions was a touchy issue. Other critics scoffed at some of the critics of such a move, some saying, "If people want to throw away their money gambling, then why not let them do it wherever they want." Many critics voiced their opinions that gambling affected more than just the players; it also had a big impact on families, friends, employment, productivity, and society in general, as well. But a pavilion for gambling was already being planned, and casino owners and other business interests attempted to block the move to bring gambling to the pavilions

It was really not surprising when gambling interests outside of the pavilion protested against having gambling in the pavilions. A number of suits were now being considered in the court system–the main allegation being that gambling in the pavilions would unfairly compete against gambling businesses in the private sector.

Mufflan would have liked to outlaw all gambling, except on a small and private basis. He viewed these activities as a way for some to enrich themselves unfairly by cheating others; in his eyes, gambling was not a moral way to earn a living. He concluded that outlawing gambling, like prohibiting many other vices, would not work as long as people wanted it. So he decided to use the pavilions as a way to control it and to reduce its negative impact on society.

At first, establishing the gambling pavilion had no great impact on existing gambling establishments and practices. There were several major differences between how gambling took place in and outside of the pavilions. Mufflan wanted to make gambling games in the pavilion fairer than outside the pavilions. He concluding that the main motivation for gambling was not accumulating wealth; except for a few gamblers, who earned a living and even became rich from gambling, most of those who won in those so-called games of chance quickly gambled away their winnings.

So there probably going to be a gambling pavilion after all–but with some major differences from those out in society. The winnings of players could not be taken out of the pavilions, and, when a player left the pavilions, his or her winnings were deposited in a pavilion's account; that money could only be spent within the pavilions– any of them. And there were various ways of spending them in any of the pavilions and the other shops, restaurants, and other places located in pavilion compounds. And at the end of each year, the net winnings of players at the gambling pavilion were heavily taxed--at the rate of least 75%--and were deducted from players' accounts and remitted to the government.

There would be some basic requirements for becoming a player at the gambling pavilion. Potential gamblers would be required to undergo background checks before they were allowed into this pavilion, to ensure that their potential losses would not hurt anyone. Compulsive gamblers would be carefully screened, and most were prohibited from visiting the gambling pavilion. Those who would be allowed into the gambling pavilion had to sign an agreement that requiring them to purchase a large life insurance policy, issued by the government, payable to the gambler's surviving family to cover any losses if gamblers got too far into debt; the consequences of extreme indebtedness racked up by players in the gambling pavilion could mean anything from being banned from the pavilions to indefinite servitude in the pavilions, which for some could be a virtual death sentence if served in the more dangerous pavilions.

Mufflan had hoped that a gambling pavilion would serve as an alternative to the usual gambling establishments out in society so that this vice would be less harmful to people and society, while meeting the urges that led to the desire to gamble. He hoped that establishing a gambling pavilion would lead to a serious discussion of gambling itself, which Mufflan saw as counterproductive and a waste of human effort and energy. Mufflan's studies showed that a relatively few gamblers actually earned a living at the tables; most of the money was earned by those who ran the gambling establishment themselves. Various states would continue to run lotteries, in lieu of inadequate attempts to finance the requirements of the government from tax revenues.

Mufflan also believed that stock trading had degenerated into an unfair gambling racket with the deck clearly stacked in favor of the large firms because those offering and marketing trading stocks and commodities often understated the risks upfront and over-hyped the amounts expected in returns. He hoped he could come up with some kind of pavilion, in which an honest stock market would operate, but that would be an enormous task and would require a lot of thinking, analysis, and research.

To Mufflan, there was a drastic need to reform the mindset of those who preyed on others for the sake of personal gain in the investment businesses. He hoped to find a way to punish those who abused business practices; perhaps through the pavilions he might find such a means: maybe some type of activities or punishment in the pavilions where dishonest traders could be punished, reformed, and rehabilitated. The most important objective that Mufflan hoped a gambling/investment pavilion would achieve was fairness: He not only wanted to make gambling and investing fairer but to also punish those who tried to violate the rules.

Mufflan's idea for the E-pavilions ("e" stood for executions) would require a lot of public relations work before they materialized. After all, capital punishment was a sensitive area–most of the civilized world having already abolished its practice in favor of long-term incarceration.

From his studies of violence, crime, and punishment, Mufflan had concluded that the execution of criminals by the state did not serve as an effective deterrent to preventing additional crimes. In effect, that process merely killed someone who had killed someone else. It was also administered unfairly: the rich and influent rarely received death sentences, most of the recipients being minorities and the poor. And, in spite of the arguments, proponents put forth to justify its use, the killing throughout society continued in society and in death chambers. Where and when would it end? It wouldn't, Mufflan concluded, but perhaps it could be reduced as a practice in society.

That conclusion led to Mufflan's idea of pavilions where executions would take place and where a more equitable form of retribution could be achieved. He didn't like to use the word justice here. Killing someone who had killed another was simply more killing— in effect, revenge. But even in the Bible–at least in the Old Testament, it was still retribution or, some might call it, revenge: an eye-for-an eye and a tooth-for-a-tooth.

Mufflan scoffed at the disparity between the U.S. calling itself a Christian nation while using the powers of the state to take the lives of other human beings. Where was the forgiveness? Instead, the current legal system seemed to be a means for getting even, and there was no way to reverse the fact that someone died, whether by murder or state-sanctioned execution.

The Execution pavilions would also be called retribution pavilions, a term which some criticized. If they ever came into being, this is how they would be operated: When someone had received a death sentence and was to be sent to the E- pavilion, the decision on how the execution would be carried out was made by the surviving relatives of the victims, if there were any, and if they chose to make such a decision. The survivors themselves were allowed to execute the condemned; they could use weapons and tools of their own choosing at the pavilions to punish the offender, or they might be allowed to provide their own. Chain saws, machetes, and acid were some of the means that could be chosen for that purpose. But the survivors would be able to sell or give away their privilege of executing the condemned to someone else.

If there were no surviving relatives, volunteers, including total strangers, would decide how to carry out the execution; people could even submit suggestions to government authorities recommending what methods to use for executions and could put themselves on a list of those who were ready and willing to administer the fatal punishment. Another way of choosing an executioner would be to sell that privilege to the highest bidder, the highest bidder earning the right to execute the condemned, or some other type of contest might be used to choose the person who would carry out the execution— in which the contestants answered questions on why they should execute so-and-so by such-and-such a method.

There were even proposals for televising the actual executions with some of the commercial revenue earned by the program sponsor going to the survivors of the victims, who had been slain. Some of the methods of execution besides the usual ones might include hanging, shooting, electrocution, poisonous gas, strangulation, burning, drowning, being drawn and quartered, being burned alive, being hacked to death with machetes and chain saws, and dismemberment by other means: acid, clubbing, being flailed alive, being torn apart and even being devoured by wild animals, being clubbed to death with sledge hammers, being run down by motor vehicles, etc.. Execution by other slow and painful methods might emerge as favorites for future executions.

Mufflan believed that a lot of people would be willing to pay large sums of money for participating in the execution of those who had been sentenced to death, even for the privilege of killing the condemned by slow and tortuous means. He also concluded that this delicate issue would require a lot of planning before any such legislation could be proposed and enacted and the E-pavilion could actually be built and operated.

Mufflan was the kind of man who did not like to see anything wasted—even the bodies of dead human beings. Since some of the activities in the pavilions would involve killing and dying, he wanted to find a means to recycle those bodies, the homicide, the gladiator, the shooting gallery, and the proposed E-pavilions being the major source of these. So why not utilize those bodies and not allow that nourishing flesh to go to waste? This matter, like the E-pavilions, was a very sensitive area, which some would call cannibalism. One thing was certain: only dead bodies would be re-cycled—some made into food products, for animals and possibly even for humans eventually—although rumors were spreading that that was happening already.

It was conceivable that one day people could be moved from place to place at much higher rates of speeds than they were presently, even those approaching the speed of light. For now conventional transportation methods would have to suffice for transporting visitors to and from the pavilions. There were already plans to construct special transportation stations, located some distance from major population centers.

At these stations, passengers to the pavilions would board the special transports that by a series of elevators, conveyors, capsules, etc. would take them to the desired pavilion compound. The final destination might be a few miles distance away, or much further. Who knows, but one day passengers might be transported thousands of miles, even to other countries, and eventually even to different planets. What was learned about experimental modes of transport to the pavilions might also be used to improve methods of transportation for communities, nations, and even beyond the earth.

Mufflan foresaw a time—sometime in the future, but certainly eventually—when pavilions might be staffed partially or totally by automated androids that would or could replace some of the human workers there. These programmable robots would be able to respond to the instructions programmed into them; they would be designed to do almost anything humans wanted them to do. They would appear as ordinary people, on the outside and in their actions, but inside they would be machines and, in place of minds, would be controlled by computer components and computer instructions. People could treat these androids roughly, even destroy them without doing permanent harm, and, after being treated in those ways, they could quickly be repaired and put back into service. Another consideration was that a droid, that is, a robot who resembled a human, didn't have any feelings that could be hurt. At the pavilions and in society, as well, people wouldn't need other people for certain activities, or would be less dependent on them, and machines would be able to do almost everything for them--almost anything.

Mufflan had foreseen a time when the pavilions would be staffed almost exclusively by machines— androids, machines capable of performing a variety of tasks. Eventually, perhaps, some of the pavilions would be totally automated. But weren't machines intended to serve people? Mufflan had asked himself. And they would be able to serve them in so many ways, including intimate ones.

In addition to programmable androids, he envisioned special kinds of beds and other furnishings which could be programmed, and, in conjunction with the instructions given to the androids, those devices and the android together would do whatever humans wanted them to do intimately. Even entire scenarios would be programmed into these automated devices, and the androids would perform or take part in them. But such innovations would take a lot of special care, since people were beings of flesh and blood; they could be hurt so easily, even killed. And they were sensitive, and their feelings could be so easily hurt.

Mufflan saw the introduction of automated androids and devices into the pavilions as taking placed in the distant future. But in the present, he believed that the world was not yet ready for such innovations. Better take it slowly, he reasoned. But it might be ready when all of the humans, who wanted to harm others and themselves, practiced their ways strictly in places like the pavilions. Then the machines could be brought in. Then humans would no longer be allowed to harm each other but could take it out on the machines, if they needed to take it out on someone.

18

EPILOGUE

Frank Gidding, married Sara Clark, and they had several kids. He was promoted to assistant editor of *The Periscope*, which he had served so loyally and courageously, and where he had met his wife-to-be, as a reporter, on an assignment in the pavilions. Earl Murphy, Frank's boss, retired after a long and distinguished career in the field of journalism.

Frank also began to write creatively—one of his most successful achievements was a novel entitled *Pavilions*. It was a bout a reporter who was tasked to infiltrate the pavilions and write up what he had observed experienced. History does, indeed, repeat itself.

Frank and his former editor collaborated on a biography of Joseph Mufflan, heavily documented with references to the articles *The Periscope* had published and from Frank's notes.

Although a number of witnesses had testified at the trial that they had not actually heard anything regarding plans to assassinate Mufflan, a number of other witnesses would substantiate that they had heard Rev. Jones say, "We've got to stop him."---a very ambiguous statement and yet suggestive of what led to violent activities and eventually to the murder of President Mufflan.

It was reported that Jones had such statements and many more to a number of his followers. One, in particular and a co-defendant, a man named Fred Wilkins, had replied, "I know how to do it; I'll take care of him." Of course, there were conceivably hundreds of possible witnesses, who could testify that he had shot and killed President Mufflan, but only a few were called to the stand at Mufflan's trial. Rev. Jones was arrested because of his association with the shooter and for encouraging violent activities.

At his trial. Rev. Jones would say that he was shocked that one of his followers would take his words to such extremes.

One commentator remarked that "Jones wanted to stop Mufflan, and, since he didn't have the guts to job the job himself, he found a lackey who would do the dirty work for him." In some news stories the term "lackey" was toned down to "follower."

Both Wilkins and Rev. Jones were tried and found guilty of the murder of President

Mufflan. Instead of being sentenced to death or long-term imprisonment, both were sentenced to the homicide pavilion for an indeterminate period of time. Some critics called their sentence to the pavilions unfair, but many supporters called it poetic justice.

As Mufflan had hoped, it was no longer necessary to sentence most offenders to death or long periods of imprisonment. Most of those, who would have received death or life sentences, were instead killed off in the pavilions or died there of natural causes. The idea of being sentenced to the pavilions as a place of punishment scared the hell of most would-be offenders and probably contributed to continuing decrease in crime rates.

Even high-minded concepts of morality could not kill off the pavilions because its founder had dealt in results—not morality. Mufflan had sought to promote a better life for his people---even if he had tried to achieve that goal in a very unorthodox way. Joseph Mufflan was dead, but the pavilions continued to operate and would expand in number and variety in what they had to offer.   They stand as his legacy, and that is the way he would have wanted it.

THE AUTHOR

Born in McKinney, Texas, in 1946, Clarence Wall's family moved to San Diego, California in 1954. In 2006, he re-located to south east Arizona.

In addition to a fulfilling career in the U.S. Army, California Army National Guard, and Army Reserves (with service in Vietnam and Europe), he worked for 20 years as a technical writer and editor. After 9/11, at the age of 55, he returned to active duty and served in Saudi Arabia.

Clarence is married, and he and his wife Ginny have a grown son and a daughter. In addition to his current work with a government contractor, he has enjoyed performing in local theater and training videos, singing at church and karaoke, playing basketball, performing stand-up comedy, and trying his hand at various types of writing.

Clarence is also the author of *Mama Squad* (available through BeachHouse Books), *Wheezer's World,* and *The World's Oldest Recruit.* All of his books are available in Kindle-versions on Amazon).

More Books!

I want morebooks!

Buy your books fast and straightforward online - at one of the world's fastest growing online book stores! Environmentally sound due to Print-on-Demand technologies.

Buy your books online at

www.get-morebooks.com

Kaufen Sie Ihre Bücher schnell und unkompliziert online – auf einer der am schnellsten wachsenden Buchhandelsplattformen weltweit!
Dank Print-On-Demand umwelt- und ressourcenschonend produziert.

Bücher schneller online kaufen

www.morebooks.de

OmniScriptum Marketing DEU GmbH
Bahnhofstr. 28
D - 66111 Saarbrücken
Telefax: +49 681 93 81 567-9

info@omniscriptum.com
www.omniscriptum.com

MIX
Papier aus verantwortungsvollen Quellen
Paper from responsible sources
FSC® C105338
FSC
www.fsc.org

Printed by Books on Demand GmbH, Norderstedt / Germany